Cambridge Elements ≡

Elements in Organizational Response to Climate Change
edited by
Aseem Prakash
University of Washington
Jennifer Hadden
University of Maryland
David Konisky
Indiana University
Matthew Potoski
UC Santa Barbara

INSIDE THE IPCC

How Assessment Practices Shape Climate Knowledge

Jessica O'Reilly
Indiana University

Mark Vardy
Kwantlen Polytechnic University

Kari De Pryck
University of Geneva

Marcela da S. Feital Benedetti
Independent Researcher

CAMBRIDGE
UNIVERSITY PRESS

CAMBRIDGE
UNIVERSITY PRESS

Shaftesbury Road, Cambridge CB2 8EA, United Kingdom

One Liberty Plaza, 20th Floor, New York, NY 10006, USA

477 Williamstown Road, Port Melbourne, VIC 3207, Australia

314–321, 3rd Floor, Plot 3, Splendor Forum, Jasola District Centre, New Delhi – 110025, India

103 Penang Road, #05–06/07, Visioncrest Commercial, Singapore 238467

Cambridge University Press is part of Cambridge University Press & Assessment, a department of the University of Cambridge.

We share the University's mission to contribute to society through the pursuit of education, learning and research at the highest international levels of excellence.

www.cambridge.org
Information on this title: www.cambridge.org/9781009559836

DOI: 10.1017/9781009559843

First published 2024

A catalogue record for this publication is available from the British Library.

ISBN 978-1-009-55983-6 Hardback
ISBN 978-1-009-55982-9 Paperback
ISSN 2753-9342 (online)
ISSN 2753-9334 (print)

Inside the IPCC

How Assessment Practices Shape Climate Knowledge

Elements in Organizational Response to Climate Change

DOI: 10.1017/9781009559843
First published online: June 2024

Jessica O'Reilly
Indiana University

Mark Vardy
Kwantlen Polytechnic University

Kari De Pryck
University of Geneva

Marcela da S. Feital Benedetti
Independent Researcher

Author for correspondence: Jessica O'Reilly, jloreill@indiana.edu

Abstract: *Inside the IPCC* explores the institution of the Intergovernmental Panel on Climate Change (IPCC) by focusing on people's experiences as authors. While the budget and overall population of an IPCC report cycle is small, its influence on public views of climate change is outsized. *Inside the IPCC* analyzes the social and human sides of IPCC report writing, as a complement to understanding the authoritative reports that underwrite policy decisions at many scales of governance. This study shows how the IPCC's social and human dimension is in fact the main strength, but also the main challenge facing the organization. By stepping back to reveal what goes into the making of climate science assessments, *Inside the IPCC* aims to help people develop a more realistic, and thus, more actionable, understanding of climate change and the solutions to deal with it. This title is also available as Open Access on Cambridge Core.

Keywords: climate, assessment, science, international governance, consensus, participation, environment, negotiations, writing, policy

ISBNs: 9781009559836 (HB), 9781009559829 (PB), 9781009559843 (OC)
ISSNs: 2753-9342 (online), 2753-9334 (print)

Contents

1 Introduction

The Intergovernmental Panel on Climate Change (IPCC) issued its first report in 1990: A definitive account of the state of expert knowledge about climate change, the scope of its impacts, and the range of solutions available. Before the IPCC's joint founding, in 1988, by the World Meteorological Organization (WMO) and the United Nations Environment Programme (UNEP), climate scientists had been publishing research about anthropogenic climate change in peer-reviewed journals, workshop reports, and assessments for government and nongovernmental science agencies, but there was no established forum for taking stock of this work *en masse* and ensuring that governments of countries around the world consider it together. The IPCC changed that. It is the premier international body for assessing climate science. And in the ensuing decades, each new iteration of IPCC reports has provided increasingly robust, certain, and alarming assessments of our warming world.

IPCC reports have tremendous authority. The intergovernmental character of the IPCC means that virtually all the world's governments adopt the same scientific reports on climate change – a necessary, if not sufficient, baseline for the political agreements meant to lower greenhouse gas (GHG) emissions and slow or manage climate impacts. The IPCC has had notable successes. Its reports helped lead to the establishment of the United Nations Framework Convention on Climate Change (UNFCCC) in 1992, which organizes international political work around climate change, and in 2007, the IPCC won the Nobel Peace Prize jointly with Vice President Al Gore for advancing climate change knowledge. The IPCC continues to shape the UNFCCC's work, including the 2015 Paris Agreement, and to inform international global governance.

Yet public and policy responses have failed to meet the urgency of the situation. We are now in a critical decade in which, as IPCC reports make clear, rapid climate action is necessary to limit global warming to well below 2°C and secure a liveable future for all. No wonder, then, that United Nations (UN) Secretary-General António Guterres referred to the IPCC's most recent Sixth Assessment Report (AR6) as "a code red for humanity" (UN, 2021), "an atlas of human suffering" (UN, 2022a), and "a litany of broken climate promises" (UN, 2022b).

How do the authors of IPCC reports understand the disconnect between the science they lay out and the action that ends up being taken – or not taken – to mitigate the climate crisis? For that matter, how do they create the reports to begin with? In this Element, we examine the inner workings of IPCC assessment to gain insight into both the scientific and the political aspects of climate change. Drawing on sustained ethnographic work conducted over the five years of the AR6 report cycle, in which we attended twelve Lead Author Meetings (LAMs) around the world and on Zoom, and conducted over 200 interviews, we explore the social and human sides of IPCC report writing, as well as the sociocultural practices that shape our understanding of climate change and of the solutions to tackle it.

From the outset, the IPCC's reports have been authored by three Working Groups (WGs): WGI, composed of mostly physical scientists, assesses the physical and natural sciences; WGII, composed of social and physical scientists, focuses on adaptation; and WGIII, composed of primarily quantitative social scientists and engineers, analyzes mitigation. The assessment report cycle ends with a Synthesis Report (SYR) that brings together a cohesive narrative about the three WG reports. This division is meant to parcel tasks out to experts from particular disciplines with particular skills, in manageable pieces, but it also embeds a key assumption into the very structure of the IPCC: The physical sciences are seen to start things off, determining the nature and scope of the climate problem, while the social sciences then add "the human element," exploring the impacts people will face and the actions we can take to solve it. But when examined from closer up, the picture becomes a little blurrier. The world doesn't fall into tidy dichotomies of science/politics, natural sciences/social sciences, or the physical climate system/human engagements with climate.

When we look behind the scenes at IPCC meetings, we can see how disciplinary norms shape the production of reports in particular ways – as well as how IPCC reports' emphasis has shifted from the physical sciences toward the social sciences over the organization's history. For example, some conventional representations of science assume that scientists can determine a "safe" amount of GHGs that can be released into the Earth's atmosphere, and that policymakers can then agree, through the rational application of this scientific knowledge, on ways to remain below that threshold. Indeed, these assumptions inform the view taken by some proponents in and of the IPCC. But in reality, climate knowledge cannot be compartmentalized in such a segmented manner.

In the sections that follow, we explore what the social and human sides of IPCC report writing look like, as a complement to understanding the authoritative reports that underwrite policy decisions at many scales of governance. By examining these dimensions, we can engage with the information produced more critically as well as more empathetically and understand what the IPCC offers to people interested in climate science, advocacy, and solutions. We can see how the social and human dimensions are in fact the main strength of the organization. By stepping back to reveal what goes into the making of climate science assessments, we aim to help people develop a more realistic, and thus – we hope – more actionable, understanding of climate change and the solutions to deal with it.

Science, Politics, Policy, and Assessment

Scientists, through their work, seek to represent nature but cannot capture all its complexity. In other words, there is no single way that the natural world can be mapped through science (Sismondo, 2010). Relatedly, the social world is contingent upon the multiple, creative, and diverse agency of the humans who are constantly inventing new ways of being, even as they are constrained by social, political, and economic structures (Shove et al., 2012). For these reasons, science should be understood as a human endeavor that is always shaped by historical and social circumstances (Shapin, 2010).

There are significant differences between how science is discussed in public and how science is carried out. For example, the IPCC gains credibility and legitimacy by claiming that it maintains a clear boundary between science and nonscience (Sundqvist et al., 2015): IPCC leaders and authors frequently invoke the axiom that they produce "policy-relevant science advice." But the lines between natural sciences and social sciences, or between nature and culture, are not universally agreed upon (Latour, 1993). Rather, they are worked out in practice. Knowing this has several important consequences for how we engage with the IPCC.

First, the IPCC is a central actor in the struggle to understand the climate problem and identify climate solutions. While originally and primarily known for detecting the influence of humans on the climate, which altered the public perception about anthropogenic climate change, the IPCC is increasingly also involved in the definition of the solutions to tackle it. In the run-up to the Paris Agreement, which was signed in 2015, the IPCC turned its focus even more toward solution-oriented assessments.

Second, the IPCC as a community is neither unified nor homogeneous. Besides being divided into WGs, Technical Support Units (TSUs), chapters, the Bureau, the Secretariat, and the Panel itself, the IPCC is a networked organization that includes both formal and informal interpersonal networks, all of which are crucial for producing policy-relevant science advice that is accurate and understandable (Pelling et al., 2008; Venturini et al., 2023). The informal work that parallels formal IPCC procedures forms a crucial social dynamic that allows the work of assessment to carry on. We were able to explore how the IPCC produces reports through both formal and informal networks and processes.

Third, because of its high visibility, the IPCC has experienced attacks on its credibility over its history, including by well-funded climate contrarians (see Oreskes and Conway, 2011). This was seen in 2009–2010 with the "Climategate" scandal precipitated through contrarians hacking the email system of the University of East Anglia, as well as some errors found in AR4. While there was much written about that episode, one of the lessons it taught is that the authority of the IPCC, or the amount of credibility and trust that the public invests in it, is not guaranteed (Hajer and Strengers 2012). Rather, the IPCC must continually shore up its credibility and perform anew its role as scientific expert. This is a difficult thing to accomplish because there isn't just one scientific method, and science is not a sure road to certainty. That is, while science is often presented in public as if it produces uncontestable or incontrovertible facts, in actuality science is a complicated process that engages with uncertainty and contingency. This is especially true when dealing with something as complex and vast as climate change, the science of which spans from hundreds of millions of years ago to projections hundreds of years in the future. This means that the IPCC frequently finds itself engaging with the plurality of unruly and contingent social and natural sciences of climate change, on the one hand, while also presenting climate science to the public as if the science of climate change were as simple and straightforward as the law of gravity, on the other hand. In this Element, we travel back and forth between the complexity of climate science and its neat and tidy representation in IPCC assessment reports.

Fourth, the political contexts into which IPCC reports are being delivered matter and are changing over time. The 2015 Paris Agreement changed the landscape within which the IPCC works by increasingly turning the focus of the IPCC toward solution-oriented assessments, shifting the frameworks through which climate change is described as a problem and how solutions are consequently posed (Beck and Mahony, 2018; Hulme, 2016). The IPCC's institutional practices require authors to avoid policy prescription, so these human dimensions of climate solutions must be assessed carefully to avoid suggestions. However, as the climate crisis has gotten worse, the science is clear about what actions have become imperative to solve the problem. Yet just because the IPCC deems some solutions possible in a technical sense doesn't necessarily mean that they're socially or politically possible.

Finally, science is situated in the broader human experience (Haraway, 1988). The origin, culture, and socialization of experts influence how they participate in the assessment of science. This then raises important ethical questions about whose knowledge counts in the IPCC and in what way. This also means that authors experience the IPCC and its assessment process very differently depending on their gender, whether they are from the Global North or the

Global South, an early career scientist or an advanced scholar, a scientist or a practitioner.

While the social location of IPCC authors influences their participation in the IPCC, scientists are also influenced by the working cultures and norms associated with scientific practices. That is, they produce scientific assessments to inform international decision-making by borrowing practices from their scientific disciplines and epistemic communities (Haas, 1992). For example, authors assess information using skills learned through peer review and laboratory and field-group interactions, and solve the problem using their familiar research methods, such as statistics or modeling. However, IPCC authors and leaders also borrow practices from diplomacy, gesturing toward the final audience for their reports as well as the milieu of international organizations in which the IPCC is entrenched. Such practices include "red team/blue team" reports (in which two groups produce oppositional reports), perspective-taking in debates, and drafting language that is politically acceptable through the appearance of neutrality. As such, IPCC authors form an epistemic technocracy, in which experts utilize the form and content of bureaucratic organizations to mediate their scientific work (De Pryck, 2021; O'Reilly, 2017).

People in the IPCC do not just passively experience the institution: They shape it, negotiate with it, critique it, and improve it, working within, alongside, and outside the formal procedures agreed upon by governments. Given this heterogeneity, we argue that the IPCC and its working components (such as chapters) do not form traditional epistemic communities – groups of experts working with and on shared sets of knowledge (Haas, 1992), even as their practices – and many of the principles guiding IPCC work – assume they do. Instead, the IPCC intentionally values diverse perspectives and seeks to introduce multiple biases and achieve agreement *through* epistemological discord. The IPCC manages this primarily through author selection, choosing authors from all the world's regions, across disciplinary fields of inquiry, and according to other subjective criteria. The IPCC authors, writing in groups, represent their foundational epistemic communities (as well as the other social worlds), co-producing climate knowledge with other authors doing the same. The collective practices of international scientific assessment lend credibility to the robust scientific consensus on anthropogenic climate change and support an equally robust international political response to solve the problem the scientists convey with such care and precision.

Methods

Our ethnographic team consists of O'Reilly, an anthropologist; Vardy and Feital, sociologists; and De Pryck, a political scientist. While much has been written about the IPCC (see, e.g., De Pryck and Hulme, 2022; Hughes, 2015), we have had a unique, sustained, and collaborative access to the organization that has allowed us to engage in original and thick description (Geertz, 1973). Research observers were not allowed into IPCC LAMs until our project.

Following an expert meeting on potential studies of IPCC processes, a special meeting at the IPCC Secretariat to construct guidelines for the Panel to consider, and consideration by governments at an IPCC plenary meeting, research access was granted to O'Reilly and her research team to study AR6 in 2018. We were not granted full access, however. We were allowed to attend the WG LAMs, but we were not given official access to the chapter and Summary for Policymakers (SPM) writing groups (though some of these groups chose to include us regardless).

We used predominantly ethnographic methods of observation and interviews to collect our data. At least two and sometimes three ethnographers attended each WG's LAM, observing the ritual of the week as it unfolded in plenary meetings, training meetings, breaks and mealtimes, and evening events. For each of the chapter writing teams that we consider as case studies, we interviewed each consenting member of the chapter at each LAM, creating an in situ oral history of each participant's experience in the IPCC at intervals in the process. We also conducted ad hoc interviews with the IPCC leaders, staff, and other chapter authors working on topics relevant to our study. After the coronavirus pandemic shut down global travel in 2020, we conducted our observations and interviews over Zoom, including observing the approval plenaries for each of the WGs. Additionally, some of us observed IPCC activities at UNFCCC meetings.

Even as we seek to introduce readers to the processes that help make scientific assessment relevant to decision-makers and the public, we note that our scholarly work often requires keeping the urgency of the climate crisis at arm's length. For example, our fieldnotes on IPCC meetings are usually quite rote, recording the formal negotiations close to verbatim alongside some general observations regarding tone, controversy, or asides. There is scholarly delight in the epistemic weeds of learning how scientific agreements and disagreements are presented in reports. But the implications of some of the discussions were often distressing. Fieldnotes from a communication meeting about crossing the 1.5°C threshold gained a tone of despair as one of us began to take on the gravity of the messages the IPCC authors grapple with communicating. Looking at a quadruple image of high-temperature projection maps, with extreme warming under every future climate scenario, she wrote, "all I see is the world on fire." Ethnographic writing – including some analytical distance – is a tool we engage to strive toward fuller description and comprehension, but our concern for the need for urgent, multidimensional, and just responses to the climate emergency drives our research.

Orientation to the Element

Section 2 describes IPCC author experiences and subjectivities, analyzing how their subject positions impact their perceptions and experiences of the IPCC process, as well as how representation matters for the quality of IPCC assessment reports. Section 3 considers WGI knowledge, demonstrating how authors navigate new and changing science in creating assessment reports. In Section 4, we detail the work undertaken by authors as they built trust and established a common vocabulary between WGI and WGII. This human effort is crucial because it can provide greater insight into how the risks of climate change are themselves changing. Section 5 interrogates the domain of WGIII, mitigation – considered a solution to anthropogenic climate change alongside adaptation. The IPCC reports generally frame climate solutions as a mix of economic policy and technological innovation, though there have been forays into cultural and behavior change in AR6. In Section 6, we analyze the relationship between governments and the IPCC, considering the malleability of expert knowledge as it moves from the authors to government approval. That the SPM is approved line-by-line by government delegates means that the text is edited to reach consensus, sometimes at the expense of clarity. However, this editing, and the intensive style of government approval, gives the text a particular kind of authority that allows the work of international climate action to proceed. In sum, we show the social complexities of IPCC assessment report writing to underscore the robust work undertaken to pro-duce the highest quality science advice in the hopes of resolving the climate crisis as quickly and equitably as possible. Many of the authors spoke about the high stakes of their IPCC work, and we hope that this commitment comes through in our ethnographic account of AR6.

2 IPCC Authors

Writing reports for the Intergovernmental Panel on Climate Change (IPCC) is a sociocultural experience shared among authors. Being selected as an IPCC author is generally considered a high-status service gig, one that offers no financial compensation. To be an IPCC author is to be identified by your nation, which nominates you, and the IPCC itself, which confirms you, as a leading climate scholar. One author noted that "my university president now knows who I am," suggesting, at a minimum, reputational benefits to the role. Authors also note the serious fun of scholarly assessment, getting to know other climate experts, having the opportunity to travel, and helping to make a difference with one of the most critical issues of our time. Authors also often publish with members of their chapter team or other people they work with in the IPCC; so along with the report citation, authors may get some publications or projects – the major expectation for university researchers' jobs – or secure new positions from the effort (Corbera et al., 2016).

Apart from the benefits, authors note the challenges of being an IPCC author – and many of these revolve around navigating the social complexities inherent in assembling experts on a topic. Academic expertise is hardly ever generalist and success often entails carving out and defending a particular perspective, making a unique contribution to one's field of study. That delineating and defense work does not translate easily into communicating more broadly about a scientific topic, balancing different views in the literature and in the author team. Gender, nationality, and career stage also impact the sociocultural dynamics of writing together, alongside personality, first language, and previous involvement with the IPCC. We observe that the experiences of IPCC authors are entangled with some of the same inequities that we see playing out in many international negotiations, particularly along the division of the Global North and South – categories that are overly simplified but nonetheless impactful at individual and collective scales.

Authors work in communities within the IPCC; they are part of chapter writing teams, Working Groups (WGs), and the IPCC itself. Each of these social formations has distinct subcultures that the leaders and people within the group construct. Each also brings together experts from similar, but also different, fields or epistemic communities (Haas, 1992). All these authors are highly trained experts, though this does not preclude hierarchical divisions within these groups. While the IPCC claims to not produce new knowledge, our previous studies disagree (Brysse et al., 2013; Oppenheimer et al., 2019; O'Reilly et al., 2012). The IPCC process and its reports synthesize and reorganize existing knowledge, drive research agendas, and inspire new research questions.

The IPCC functions, but it doesn't always function smoothly or in the same way for every author. Group practices established with the intention of facilitating diverse forms of communication and a full airing of expert views can fall by the wayside as deadlines approach – or pandemics emerge. Writing IPCC reports is hard, serious work, especially given that the authors are volunteers. This section explores some of the shared challenges in the author experience, followed by some additional, systemic barriers that confront authors. If representational diversity is important to producing climate science assessments of the highest quality, as IPCC claims, the organization is just getting started with creating the conditions that permit all authors to substantively participate.

Introducing IPCC Authors

One of the first tasks for the IPCC is to select its authors. Selecting IPCC authors is a negotiation in which the leadership of the organization seeks expertise alongside representational diversity in categories such as nationality, gender, and career stage. This representation is essential for governmental approval and buy-in of the report; governments demand that their country and region are well represented in the documents and in the process. This representational diversity also serves as an inoculation against critiques of bias (Oppenheimer et al., 2019). If there are authors covering most of the world's regions, and an attempt is made to reach gender parity, this thinking goes, the resulting assessment produces a "balance of bias" that is more objective for having included so many perspectives (Boykoff and Boykoff, 2004). This is also referred to as "strong objectivity" (Harding, 1995).

But as anyone who has held a marginalized position in a meeting knows, just because a person is in a room – representing, say, "women," or a "small island developing state" – does not mean that their views receive equal consideration or equal treatment in the final text. Long-standing cultural and gendered communication styles result in some experts yielding their positions while others expect theirs to be yielded to. This means that some authors must work much harder to communicate: in their second or third language, from nations and institutions with less power, and in the context of their other responsibilities and marginalized identities, including being constantly asked to represent the "diversity slot" in climate science, with little to no collegial or institutional support.

The IPCC Bureau – the chair, vice chairs, WG/Task Force co-chairs and WG vice-chairs – and the Technical Support Units (TSUs) select authors from the pool of experts nominated by governments and observer organizations, and sometimes from their personal networks. Each chapter has two, or occasionally three, Coordinating Lead Authors (CLAs) who manage the chapter-level writing. There is always at least one Global North and one Global South CLA for each chapter, and there are usually some power imbalances built in, in areas including funding, English language skills, and integration with scientific communities. Under the CLAs are the Lead Authors (LAs), who are usually responsible for sections of expertise in the chapter. They can count on the support of Chapter Scientists (usually early-career scientists) to accomplish the immense yet often under-acknowledged work of compiling various documents, keeping them updated, and loading them onto the IPCC's secure platform. Chapter teams can invite Contributing Authors (CAs) to provide parts of the text in their areas of expertise if those areas are not sufficiently covered by the CLAs or LAs. Contributing Authors do not attend Lead Author Meetings

(LAMs). Finally, Review Editors (REs) ensure that the work performed by the authors is in line with IPCC practices and procedures. Most author travel is paid by authors' respective national governments. Authors from developing countries and economies in transition often have some of their travel expenses covered by the IPCC Trust Fund, which is mainly funded by voluntary donations from wealthier governments.

The role of the CLAs is crucial. There is a long list of responsibilities that they take on, and CLAs undergo IPCC training and additional meetings at LAMs and online. At a minimum, they must coordinate the input from LAs working on their chapter, and they must coordinate with CLAs of other chapters to decide on questions of specific content. Most of them participate in the writing of the Summary for Policymakers (SPM). But beyond this, CLAs must also – to greater or lesser degrees – practice leadership and conflict resolution. In our interviews and observations, we heard and saw how there are significant differences between chapters; some chapter teams worked together efficiently and amicably, while others experienced more tensions, and sometimes rancor – to the extent that they were described as "dysfunctional."

Coordinating Lead Authors, of course, have different leadership styles and even differing expectations of their role. One experienced CLA, who had held the role for multiple cycles, emphasized that he did not write – the LAs wrote and the CLA provided comments, edited, and synthesized the entire chapter. On the other side of the spectrum, the CLAs in a WGI chapter ended up taking on much of the writing when their LAs did not deliver their work on time. Other CLAs managed more collaborative approaches to team writing, with mixed results. Coordinating Lead Authors also have different styles of managing conflict; like all other subjective skills, this is influenced by culture, language, gender, and even their disciplinary training, as "how to do science" varies among fields.

Each WG report contains 12–18 chapters, and each chapter has 10–15 authors, sometimes more. In total, 723 experts contributed to the IPCC's Sixth Assessment Report (AR6) – 233 to WGI, 262 to WGII, and 228 to WGIII (Tandon, 2023).

Orientation to the Lead Author Meetings

Each WG holds four-week-long LAMs. For AR6, the LAMs took place at the following times and locations (Table 1):

LAMs are invaluable week-long meetings for interacting in person, building relationships, and coordinating within and between chapters. The LAMs introduce, at intervals, layers of assessment techniques and built-in goalposts to manage the massive writing process and its various reviews. In plenaries,

Table 1 Location and dates of the AR6 WG LAMs

	WGI	WGII	WGIII
LAM1	Guangzhou, China June 2018	Durban, South Africa January 2019	Edinburgh, UK April 2019
LAM2	Vancouver, Canada January 2019	Kathmandu, Nepal July 2019	Delhi, India October 2019
LAM3	Toulouse, France August 2019	Faro, Portugal January 2020	Online April 2020
LAM4	Online February 2021	Online March 2021	Online April 2021

authors are socialized into using IPCC-specific language, the differences between assessments and literature reviews, the expert and government review processes, and intercultural communication, among other things. The LAMs roughly correspond with the mandated drafts: The zero-order draft (ZOD), an informal document reviewed by experts and authors internal to the IPCC; the first-order draft (FOD), which undergoes expert review; the second-order draft (SOD), which undergoes expert and government review; and the final draft, which goes to the approval plenary for adoption.

The LAMs are a flurry of activity from morning well into the evening. The official schedule consists primarily of plenary meetings, which all authors attend, and dedicated time slots for each chapter writing team to work together. There are several other meeting types, such as the meetings for cross-chapter papers and issues, and for the SPM. In addition to officially scheduled items, the LAMs also offer authors many informal opportunities to work together on ad hoc issues as they arise. Often, they will meet over breakfast, coffee breaks, lunch, dinner, or in the evenings.

Authors perform scientific assessment through a series of interconnected activities, practices, and conversations that span multiple interpersonal networks and groups. An IPCC author needs to know how their contributions fit together with those of other authors in their chapter; at the very least, clear communication is required to work efficiently within the chapter group. In addition, many authors also contribute to chapters outside of their own through complex sets of conversations convened in breakout groups (meant to solve puzzling issues), cross-chapter boxes and papers (devices to communicate across chapters), cross-WG boxes, and glossary definitions. Indeed, for AR6, a considerable amount of work that took place at the LAMs was the work of coordinating with authors in other chapters or WGs.

We saw authors meet several times over a period of years, during which time authors' interpretation of the IPCC and their role in it often changed. We also observed how group dynamics played out for different chapter writing teams. We observed the complicated, rewarding, uncomfortable, exciting, tiring, sometimes tedious, sometimes frustrating, and occasionally joyous process of being an IPCC author. From this perspective, the IPCC is much more than a series of protocols or formulaic instructions. It is a collection of people who are, to greater or lesser degrees of success, trying to work together collectively to produce knowledge of climate change. As can be expected from such an endeavor, authors had many different experiences – some became disenchanted with the process, while others were satisfied. Some even quit.

Author Experiences

In interviews, we asked what made a good author. As might be expected, most people talked about the importance of having scientific expertise in a particular field or subject area. For example, one author told us, "the expertise you have has to match some of the expertise that is needed in the chapter where you are specifically." Similarly, another author said, "So all of us have a good sense of where the literature is, because we've been part of it. Even me, I'm one of the younger ones here . . . . But I know the field. And that's the same for everybody, we're all experts in our own way."

After mentioning expertise, many authors went on to speak about the need for authors to have what are sometimes referred to as "soft skills," or the ability to listen and to communicate respectfully while also making your own opinion heard. For example, a CLA told us,

> You've got to be a team player. . . . I mean here, we come here all volunteers. That's very, very important. We are not paid to do the IPCC job. . . . And you need to have a good heart, to be open-minded, and to accept different views.

For new IPCC authors, this may be the first time that they have participated in any kind of scientific assessment. All the foregoing factors mean that chapters are the first time that many authors have had to collaborate with others with whom they share few similarities. Despite all this difference, chapter writing teams must somehow come together.

Each author we interviewed spoke about the challenges of workload. IPCC authors are already busy people – most of them are active researchers or practitioners – and they have lives outside of being an author.

Some authors, especially from the USA and Europe, sometimes (though not always) receive teaching releases or a reduced service load while writing IPCC reports, but people from institutions that do not have the capacity to help their faculty in that way write the IPCC reports while also carrying their everyday workload. As an LA told us, "the challenge of sort of trying to balance this with a very busy job and busy family life" is an ongoing issue. Another recalled, "what is challenging is that nobody has time. Everybody has to do it in between [other things] and everybody has time at different moments." The workload fluctuates depending on several factors, including deadlines for the chapter and the contributions of other authors. They also must consider the timelines of related processes external to the chapter. For example, one WGI CLA told us that their chapter tried "to be a little more aggressive at the beginning so that will give us more room later on if something happens." That strategy would give them some more leeway if, closer to the time that the final draft was due, new literature emerged that they needed to include, or if the review process highlighted issues they had overlooked. In addition, several authors commented on trying to manage the overlapping timelines of the three Special Reports (SRs) that were also part of the AR6 cycle. As a CLA told us, "I was also working on the 1.5 report at the same time [as the WGI report], so the timing was really not optimal because we had the approval session [for SR1.5] at the same time as we had to basically finish this zero-order draft."

The workload was often unevenly distributed among authors. Many interviewees told us that not all authors contributed equally to the writing of their chapter. Some became disengaged over the process; some never really engaged in the first place. Others (CAs, TSU members, or Chapter Scientists) stepped in, but were not always given full credit for their contribution. A CLA recalled:

> You get all these people who just sign up to it, who then don't do any work. We had people in our chapter who did zero work. And then we have people who, as I say, are very unpleasant to other LAs, who do very small amounts of work and are very poor team people. Because they're all these well-known people who have got to the point where they've got like X postdocs who just do all the work for them. ... We have many CAs in our chapter also because our LAs just didn't do any work or did very little work or, you know, had all sorts of awful things happen to them ... ... We have many, many CAs who have done more work than our LAs.

Life also happened while the IPCC cycle was underway. Authors change or lose jobs, move from one country to another, fall ill, have children, get married, or get divorced. Perhaps one of the most upsetting reminders of this kind was the death of Gemma Narisma, a WGI CLA, in March 2021. Her chapter, the WGI

Atlas is dedicated to her memory. Authors were also impacted by political circumstances. For example, some authors who worked for the US government were unable to attend WGI LAM2 in Vancouver because the US government shut down due to Congressional intransigence on budget issues. An American climate modeler told us, "The Department of [redacted] won't pay for me to be part of IPCC and so we had to get separate University of [redacted] funding from the office of [the university] President." Another American author, Patrick Gonzalez, was the science advisor to the National Park Service. However, the Trump Administration did not support his participation in the IPCC, refusing to pay for his travel and sending him a cease-and-desist letter regarding his Congressional testimony about climate change in National Parks as well as his IPCC work (Plumer and Davenport, 2019). Dr. Gonzalez insisted on participating in the IPCC anyways, paying his travel expenses out of pocket and under his affiliation with the University of California at Berkeley.

Minoritized Author Experiences

The term "minoritized" is increasingly used to refer to social groups and epistemic perspectives that may be disadvantaged, marginalized, or treated as minorities, even if they represent numerical majorities in a given context (Gunaratnam, 2003). When examining the practices of the IPCC and their outcomes, it is essential to consider issues of gender, experience, ethnicity, and linguistic ability, which can create inequalities in participation. While many people's names may appear on author lists, the contributions of minoritized individuals do not always get into the report in the same way as those of more advantaged authors. This came up in many interviews.

Many authors who were women and/or from the Global South felt they had to work harder to get their perspectives across. An author noted how minoritized authors were treated, in her view:

> you got that sense you need to be proving yourself, you needed to show why you were there to some people. Sometimes you got the sense of artificiality, or the politeness of inclusion that is not really helpful. I think it is something crucial that everyone should know about. It's like the gender issue, right? Sometimes you see men treating women generally nice, but it's not because they are really thinking about it, it's just because it's like, "Oh, it's politically incorrect to say it like this." But in fact, their mind or their behavior is doing something completely different.

Another female author from South America described the accumulating factors that make it more challenging for some authors than for others to participate in the IPCC:

> It was challenging because I do not speak English. English is my third
> language and sometimes you are like "How do I say this?" It was challenging
> because I am a woman. This work is male dominated. And when I joined,
> I was really young. There were all these hierarchy issues and it was hard for
> me to develop a thick skin to say, "I don't care. . . . I will get my point across.
> I will navigate these without getting bitter or persuaded." . . . It is hidden but
> it's determining who gets to say what and who gets to listen, right?

English being the working language of the IPCC, non-native speakers also dealt with the mental burden of linguistic translation alongside technical thinking. Authors and their writing teams sometimes found ways to decrease this burden. For example, some non-English-speaking authors chose to share their views in writing, rather than participating in discussions in real time.

Another issue is the fraught relationships that sometimes exist between academic and nonacademic authors. While the IPCC has sought to be more open to practitioners, because their practical knowledge is important to understand enablers and barriers to climate action, they often struggle to find their place in a process that remains dominated by academic norms. Authors whose institutional affiliation is not considered prestigious enough may also face condescension. As an interviewee noted,

> If you do not have a good affiliation, you are nothing, you do not exist in the
> IPCC . . . . People told me once I left [my job] that I should not be in the IPCC,
> because I did not have an affiliation. . . . It is violent. So now that I have
> become aware of that, I try to find myself good affiliations.

Some authors experienced greater challenges than others in navigating their lives alongside their work. For example, some authors were parents of young children, and some of those needed to bring their children to the LAMs. Of all the LAMs, only the meeting in Faro, Portugal, provided on-site childcare. Even when childcare was provided, parent authors missed some work that took place in informal settings during the evening. As one interviewee told us,

> a person with a child who's at one of these meetings, it might be more difficult for
> them to either attend or get invited to a meeting that's going to happen at eight
> o'clock at night in the hotel bar to talk about one thing or another kind of
> thing. . . . I think that those kinds of informal relationships really do make
> a difference.

As caregiving responsibilities typically fell more squarely onto women, that meant that informal gatherings and quickly arranged meetings or those that ran over time were more likely to exclude women and therefore emphasize men's contributions.

Authors from low-income countries face similar difficulties. For example, these authors may find it necessary to stay in less expensive hotels farther away from the meeting venue instead of those on the IPCC-provided accommodation list. They thus have fewer opportunities to connect with other IPCC authors, for example over breakfast in the morning or in the hotel lounge in the evening. Authors staying in these hotels can easily connect with one another, but it is much harder for them to establish ties with better-funded researchers. These kinds of informal networks can be crucial for establishing good working relationships in the IPCC. Therefore, authors from least-developed countries often face difficult choices about whether they should spend their personal savings to advance their research and career goals.

Getting visas to travel is an unequal experience, with the heavier burden usually borne by travelers from the Global South. One author described her situation in getting from her African home country to Delhi, along with one of her colleagues, in detail:

> There are no embassies in my state, so I have to travel, fly actually by air to [capital] or [other city] to do these kinds of processes. So it takes a lot of financial commitment and time, which sometimes I do not have ... . A colleague of mine from the same university who managed to get the visa, now couldn't fly directly from [home city] to Delhi. He flew halfway, he got to Frankfurt and he had to be turned back. He didn't board the next plane, because he didn't have a visa, a transit visa. So these are some of the processes, the problems we go through. You need to get a transit visa in addition to the visa you already got. And this takes days, sometimes weeks, to process, not talking about time and financial implications.

The move online during the pandemic often exacerbated inequalities. While some felt that Zoom meetings leveled the playing field for low-income country authors (for instance, by removing the burden of having to obtain visas), many faced technical issues that threw up another barrier to participation. The author above described her experience:

> I had challenges, making sure I had a wifi to attend the meeting, stable wifi, and most of the time it was so bad, I couldn't talk, I could manage to join ... and so, I went the extra mile because since I didn't attend Delhi, I made sure I didn't miss this. I discovered that because I missed Delhi, it was a bit difficult for me to catch up with the process, the writing process of the group .... And I recall, the meetings were really long, especially like the plenaries, they were really long. And you can imagine the amount of megabytes and everything that was just running like this. Every other day, I had to keep loading my phone data. We are in a lockdown where I stay because of COVID-19. So I didn't have access to get wifi in the house. So I just had to use my phone. And that one, I could buy directly into my phone, from the bank. I could buy my data and use. So, that's how I was able to attend the eLAM.

Another author reflected at the end of the assessment report cycle that most of
the key contributors to the SPM were European or American, stating: "I don't
know how many people from the developing country participants were actually
engaged with the SPM, for example, because, you know, a lot of people struggle
with remote connection, big time, you know?"

Authors are aware of cultural differences and inequalities among themselves.
One CLA explained how they tried to encourage different forms of contributions so that everyone felt comfortable participating:

> Being a developing country expert and living and researching on developing
> countries for a very long time, I also try to see how the developing country
> perspective also gets into the chapter because we all live in our own worlds.
> So, we have expertise of our own worlds and, culturally, the developing
> country and developed country works differently, right? The work culture
> is also different . . . . We are writing a global report, but we are also sensitive
> to cultural differences, how people work, how people interact, and how
> people take leadership. Some people like to write more, but the language
> barrier is also there, but then they have very good ideas so making them write
> rather than speak . . . sometimes people do not like to speak in the group, but
> then if you talk one-on-one, then you really get good things. Sometimes some
> people just take over the meeting, right? . . . We need to make it more clear
> that everybody needs to come up and voice their parts.

Finally, many new authors, and in particular early career scientists, faced
challenges in acclimatizing to IPCC procedures and practices, meaningfully
weighting in on the assessment process, or having their role properly recog-
nized. A Chapter Scientist, whose main task ended up being checking his
chapter's references, noted with hindsight that "it is not very rewarding . . . . It
is a lot of work" – while also recognizing "that you gain from being exposed to
the IPCC process and that you can do networking." In another case, a Chapter
Scientist ended up writing a big chunk of their chapter but was not promoted to
an LA position.

The contribution of early- and mid-career scientists is generally welcomed, but
it can sometimes create tensions. An author notes that the changing demographics
of IPCC reports are a challenge to some – but results in better assessment:

> There's been a sense this time of a lot of early to mid-career new authors
> coming in. There's sort of a younger feel. Less North America. More devel-
> oping countries. And I think some of the senior scientists from developed
> countries are possibly finding that cultural shift a challenge because we've
> had at least two incidents this week of senior scientists effectively diminish-
> ing the expertise of more junior scientists—of which three of the four people
> involved were women as well. So I think the idea that early to mid-career
> scientists can be experts and be possibly even better at assessing the literature

because we may have more time, we may feel anxious about doing a good job so we're incredibly thorough whereas a professor might have more, you know—in the UK a professor's really the top of the field and may have more administrative managerial responsibilities, may think that they know the literature without going into it in as much detail. I think there are studies that show this is the case for peer review, that early to mid-career scientists can be more thorough in their reviewing. So I think there needs to be a bit of a shift of what an expert and a good IPCC author is because it isn't necessarily just about all the professors in Europe and North America, but it's something more diverse than that.

Authors contend with their everyday lives and their subject positions in these diverse groups, performing intense climate assessment work as volunteers. They generally do it, despite the challenges, with a high degree of cooperation and a strong sense of purpose: Most IPCC authors know this work is important and that their reports help decision-makers establish climate policies that can respond appropriately to the scale of the problem.

IPCC Responses

Bureau members are in charge of ensuring that the assessment goes smoothly and that authors feel respected. These leaders are almost always past IPCC authors and carry their experiences with them into their new roles: They are not blind to issues of unequal treatment, but some are more sensitive to them than others.

The IPCC has enacted numerous initiatives to facilitate the social side of scientific assessment. While imperfect and with plenty of opportunity for more and better work to address inequalities in the assessment space, many IPCC leaders and authors approach the social dimensions of global climate assessment as something critically needed to produce the most accurate and representative reports possible.

Regarding geographic representation, IPCC leadership has sought to increase the participation of authors from the Global South, with mixed results. In AR6, they accounted for only 35% of the bulk of authors, compared to 31% in AR5 (Standring and Lidskog, 2021). Working Group I greatly increased its proportion of authors from the Global South (from 21% in AR5 to 33% in AR6) but remained the least geographically diverse WG – WGII and WGIII both reached 36%.

Regarding gender distribution, significant improvements have been noted, with women constituting 34% of authors in AR6 (vs. 21% in AR5). In 2018, the IPCC established a Task Group on Gender to address gender-related issues within the IPCC. It published a comment in the journal *Nature* that reported findings from a survey they sent to over 1,500 IPCC authors (Liverman et al., 2022). The article draws attention not only to barriers that prevent women's full and effective participation in the IPCC but also to how these barriers can be compounded by other structural issues:

> The survey highlighted the importance of other dimensions of diversity that intersect with gender . . . . Several respondents reported seeing themselves or colleagues be brushed aside owing to a lack of fluency in English, or to youth, race, gender or being from developing countries. (Liverman et al., 2022, p. 32)

In 2020, the IPCC adopted a Gender Policy and Implementation Plan. The plan's vision statement notes:

> The IPCC is dedicated to pursuing a future state where gender is mainstreamed into its processes in an inclusive and respectful manner and where there is gender balance in participation and where all have equal opportunity irrespective of gender. In so doing, the IPCC will raise awareness of the benefits of gender equality. (IPCC, 2020, p. 2)

Many IPCC leaders worked hard to create a change in the culture of the WGs, and some took strong action to provide solutions. A good example of this kind of culture change comes from an incident reported by an early-career female author. In her chapter group, another author had bemoaned that their team lacked expertise on the topic that was her specialty. When she noted this to the author, he grilled her about who her advisor was, implying that she lacked not only the subject matter expertise but also the academic pedigree to claim it. When she lodged a complaint with a WG co-chair, the co-chair immediately addressed this issue in the full plenary meeting of close to 200 authors. The co-chair reminded the authors that it was essential that all authors take responsibility to enable full participation from everyone. This was not an empty gesture. To demonstrate their commitment toward enacting greater inclusion, the TSU contracted an organization that specializes in facilitating communication to provide training for the second LAM in Vancouver. The organization addressed the plenary, and offered a series of workshops, drop-in sessions, and one-on-one consultations on the topic of inclusive communication. The TSU also hired them to attend LAM3 in France. At both LAMs, the co-chairs drew attention to these workshops and opportunities during plenaries. The co-chairs also made a point of highlighting collective efforts to value diversity, give time and respect to different voices, and continue to improve communication among authors. An author from the Global South told us that the workshops were helpful in terms of encouraging non-native English speakers to participate. "You know, sometimes we feel that we do not have the confidence to speak up. And by attending the [training], I feel that is very helpful for us."

These workshops and training sessions did not go far enough for some people, however. Some authors commented how, because the sessions were voluntary, those who are most in need of the training did not attend. From the contrary perspective, the author who complained about the lack of expertise in the situation above angrily described how such initiatives were not needed and that authors should "learn to be adults." He quit after that LAM. Still others noted that they already were familiar with the content of the training, and that the IPCC leadership did not set specific goals for what was to be accomplished from it. Also, because the workshops were voluntary and scheduled at the end of the busy meeting day, they seemed low priority and, as one author put it, "tokenistic."

While the incident at WGI LAM1 spurred action in the form of inclusive communication training, other incidents did not, either because they were not reported or because the leadership did not intervene. In at least one case, the leadership itself was the problem. For example, an interviewee told us how a group of authors went on strike for several months by refusing to work, following the unfair dismissal of a TSU member. They resumed working on the condition that the individual responsible for the firing would not interfere in the writing of the report.

The IPCC's efforts on gender and inclusion show consideration of authors' concerns and some attempts to make participation more equitable across differences. Authors tend to be aware of these issues as well and have become less hesitant to bring them up. Many noted that AR6 was far more inclusive than AR5. But more efforts will be needed, especially coming from the leadership of the IPCC.

Conclusions

Communities do not emerge fully formed, out of thin air: People make, uphold, and change them through their participation. A functional chapter writing group, where people (and their expertise and ideas) are fully considered and respected, was the most crucial indicator of author satisfaction with the IPCC, according to our interviews. However, even with the best intentions, the expected stresses of deadlines, paired with unexpected stresses like COVID-19, meant that well-intentioned group practices were sometimes left by the wayside, if they were established at all. Due to long-standing professional and domestic arrangements that have long favored them, we observed that white, more established, European/American men's views often became over-represented in the final report while more socially marginalized authors were more likely to give up pushing to have their views included, due to other responsibilities that preclude them from putting up the fight required to ensure their representation. As the authors explained to us, who is on the chapter team matters, but that's only the beginning: How their perspectives become part of deliberations and part of the report itself makes a difference for producing truly inclusive and integrated climate knowledge. These sociocultural experiences shape the knowledge produced by the IPCC, making some types of knowledge more visible and others less so. These experiences, finally, also shape the level of authors' buy-in of their chapter, with some expressing stronger ownership of the chapter's conclusions and others less so.

3 Assessing Climate Science: Making and Communicating Projections

Authorship in the Intergovernmental Panel on Climate Change (IPCC) requires working with a plurality of sociocultural differences, as we showed in the previous section. At the same time, authors must also work with huge quantities of complex and sometimes incompatible scientific data. They have to judge the quality and agreement of scientific information and find ways to communicate it clearly to policymakers. In this section, we detail one way that IPCC authors have tried to accomplish this: through the 1.5°C and 2°C temperature targets. Exploring this topic allows us to show how IPCC authors face not only scientific and technical but also communication challenges. IPCC authors invest considerable effort in shaping narratives in their reports. But once they are published, these messages take on social lives of their own as they are picked up by activists and others who urge politicians to act. This section examines the origins of the idea of limiting warming to 1.5°C, followed by an analysis of some of the scientific maneuvers IPCC authors used to create and communicate robust temperature projections. Scientists are engaged in a difficult, collective balancing act of precision and politics, as they seek to underscore the urgency the climate crisis demands without giving up their own professional credibility, or that of the IPCC.

Origins of the Temperature Threshold and the Special Report

For decades, climate scientists and politicians used a 2°C temperature increase to designate the threshold between manageable and dangerous climate change. The 2°C increase became known as the upper limit the Earth could tolerate before we started to experience the worst impacts of climate change (Randalls, 2010). However, as early as 1992 when the UNFCCC was being formed, a small contingent of scientists and policymakers noted that a world warmed to 2°C would be devastating to ecologically and politically marginal communities, places, and states. Thanks largely to the work of the Alliance of Small Island

States (AOSIS), limiting warming instead to 1.5°C became the goal for people concerned with climate justice. In 2015, in a diplomatic win for AOSIS and allies, the signatories of the Paris Agreement agreed to enshrine a 1.5°C limit as part of its long-term global goal (Cointe and Guillemot, 2023).

During COP21, the IPCC and UNFCCC held a joint side event on planning for AR6. As we entered the full room, we grabbed headphones – not for interpretation, but because the echoing, cavernous space of the Le Bourget conference center made it necessary to speak through mics and listen through headphones. Otherwise, the growling background murmur of events on the other side of the thin, temporary walls made it difficult to differentiate the noise from the words of the event speakers, discussing the plans for the upcoming topics for the special reports.

At this side event, most of the conversation focused on a pending decision for a proposed IPCC Special Report on limiting global warming to 1.5°C (IPCC, 2018). The room – mostly populated both onstage and in the audience with IPCC-affiliated authors – buzzed with concern over this idea, for two key reasons. First, the Special Report topic was "invited" by the UNFCCC. These invitations are the sort one cannot refuse, and the unprecedented nature of the request for scientific knowledge coming from the policymakers raised some red flags about the IPCC's institutional independence and scientific integrity. Second, some outspoken experts in the room were unsure that such an assessment was possible: The relevant scientific knowledge did not exist, as such, in the peer-reviewed literature.

This was due, in large part, to the fact that the climate science and policy worlds had revolved around a 2°C target for much of the history of both the IPCC and UNFCCC. While the original Framework Convention's main objective is to "prevent dangerous interference with the climate system" without a number attached, the 2°C target had gathered so much political and scientific momentum that it made some experts stop in their tracks to try to imagine a lower target. Would there be enough peer-reviewed scientific literature in place to assess the risks, impacts, and implications of a 1.5°C warmed world? If not, their task seemed nearly impossible.

Some of the audience and panelists argued about where 2°C had come from in the first place. The policymakers seemed to be operating under the assumption that the number had come "from science." Somehow, in the mix, 2°C came to represent a generally safe operating space to which global temperatures could rise without too much impact. That newer research was showing impacts at lower temperature thresholds was a concerning surprise.

Some scientists present noted that 2°C was a political number, negotiated in the light of research, but also with an eye toward political and economic feasibility. What did a future look like that transitioned fossil fuel energy production and consumption while balancing economic considerations? With vast inequality among countries at various stages of development, an international agreement requires attention to national interests all along the economic spectrum. Two degrees, they reminded one another, had always been a compromise between science and politics.

Nonetheless, climate advocates had realized that a world warmed to 2°C was not a world that avoided climate impacts, particularly in already marginal environments, which also tended to be the homes of politically marginalized people. There was a push from climate justice advocates, with the backing of the powerful bloc called the High Ambition Coalition (founded by the Marshall Islands with members from least developed to most developed countries, including the USA), to limit warming to 1.5°C in the Paris Agreement, being negotiated in nearby rooms. People at the UNFCCC wanted to understand the scientific differences and impacts of limiting global warming to 1.5°C in comparison to 2°C. Discussions about 1.5°C did not come from nowhere. Reporters of the Earth Negotiations Bulletin note a longer history of the 1.5° C focus:

> The roots of a 1.5°C target stretch back further than that, to research commissioned by the Alliance of Small Island States in 2008, and subsequent pressure from those states and environmentalists, that helped realize the first mention of the 1.5°C target in the UNFCCC's 2010 Cancun Agreements. (2018, p. 19)

While the request to the IPCC came from the UNFCCC, the interest in the lower temperature target comes from vulnerable, often developing, and geographically and politically marginalized people and places.

Some climate scientists suggested the impossibility of holding the temperature increase to 1.5°C; even meeting 2°C challenges societies, political will, and the pace of economic and technological transformation. Taking up a multiyear assessment on the issue was a fool's errand, they said, but they went along with the decision to accept the invitation from the UNFCCC. It allowed them to conduct new research; it was a pressing request from an important institution; and it was a way to be responsive to the concerns of people in climate-vulnerable states and places (Guillemot, 2017).

In the end, the IPCC got to work on the Special Report on Global Warming of 1.5°C (SR1.5), moving with surprising speed from invitation in 2015 to publication in 2018. Along with scheduling a scoping meeting early on, the IPCC publicized calls encouraging researchers to study and publish their findings on the topic ahead of their publication deadline, so that their work could be assessed (see also Livingston and Rammukainen, 2020). The IPCC report publication deadlines regularly drive much of the climate science research and publication machinery, but it was rare to see this push for research from the outset. The Special Report was altering the familiar modeling terrain that kept 2°C as a lower limit (O'Reilly et al., 2012). These changes from SR1.5 reverberated throughout the Sixth Assessment Report (AR6) cycle, as authors continued to align their temperature assessments with the new frameworks created in the Special Report.

Additionally, SR1.5 mapped the notion of carbon budget – itself a complex way of representing how much CO_2 can be added to the Earth's atmosphere without exceeding a temperature threshold – onto time. Specifically, the report estimated that the world would cross the 1.5°C threshold somewhere between 2030 and 2052. Activists picked up on the 2030 number as a "deadline," and the countdown began immediately: As of 2018, activists declared, we had "twelve years left." Several IPCC leaders and authors – many of them well-versed in interacting with governments and communications strategists – have expressed frustration with this message, both publicly and in our interviews. They pointed out that, first, there were not twelve years left to act if we wanted to limit warming to 1.5°C; the time to act was immediately. And, second, while the timescale might help people comprehend the urgency of the issue, it also suggests that the global climate system will move, in a linear or stepwise fashion, toward a threshold that we will cross, or not cross, at the projected time. In fact, given the complexity of the global climate system, anthropogenic climate change is not unfolding in a linear way that we can, at some point, simply and immediately stop.

Some scholars, including IPCC authors, expressed frustration with IPCC messaging, which they thought should more clearly counter the notion of a twelve-year deadline, to preserve the credibility of both the IPCC and climate science wholesale (Asayama et al., 2019). While the report garnered more media attention than any other IPCC report, its authors feared that this attention came at the expense of losing some of the precise messaging the IPCC is known for (Boykoff and Pearman, 2019). Indeed, during AR6, IPCC authors told us that the twelve-year deadline was a runaway message that they anticipated would bring them continued communication challenges, since an increase of global average temperature of 1.5°C in any one year does not mean that the threshold has been breached. How the 1.5°C threshold has been communicated in the media, however, makes it seem as if determining when the threshold has been crossed is relatively simple. This reduction of complexity, coupled with the likelihood for global temperature increase for any one year to be above 1.5°C, led one author to predict that "all hell will break loose" if one year does top 1.5°C, because of the passionate, engaged, and enthusiastic climate activism that has coalesced around this number. Recognizing this, IPCC authors do not want activists to give up when the threshold is crossed. Even if 1.5°C is breached, limiting warming as much as possible matters. As climate researcher Zeke Hausfather noted in an interview, "the future doesn't boil down to either a 1.5°C future or an apocalyptic hellscape. Climate change is ultimately a matter of degrees rather than thresholds" (Bassetti, 2022).

Authors approached the challenge by trying to determine how best to communicate the probabilities and thresholds in advance, to use as reference points when the time came. A single year crossing 1.5°C does not mean that the threshold has been irreversibly breached: Annual global temperatures could wobble above and below the line for quite some time, making the permanent crossing detectable only in hindsight. For the IPCC, in order to be able to say that the threshold has been crossed, the average temperatures over twenty years should be 1.5°C higher than the average of the twentieth century. But as we show below, determining how to measure this is an extraordinarily complex endeavor – not to mention having to wait twenty years to find out it already happened. In the following three sections, we discuss the effort in AR6 to reconcile different ways of measuring global temperature, to estimate the carbon budget itself, and finally to build the scenarios and pathways the IPCC used to project global warming. In all three of these sections, we show how the process of marking a threshold between safe and dangerous climate change is fraught with difficulty.

Making Global Temperature Coherent

During AR6, the IPCC changed their standard for expressing global temperature. Or, rather, they implemented a standard after vacillating between two measures of global temperature in the previous assessment reports: global mean surface temperature (GMST) and global surface air temperature (GSAT). Typically, calculating observed, contemporary global temperatures has involved combining sea surface temperature with temperatures over continents to arrive at GMST. Conversely, models that project future temperatures use GSAT, which is air temperature above ocean and land at a set elevation. Scientists had assumed – until recently – that the two measures were roughly equivalent.

By the time of the second WGI Lead Author Meeting (LAM2) in Vancouver, there was a proposal circulating to use GSAT instead of GMST across the entire report: This was eventually accepted and implemented. Working Group III followed suit to make its scenarios legible alongside the WGI science. But this created a potential problem. While GSAT was considered more accurate, observed warming using GSAT also resulted in higher temperature increases compared to GMST. The authors were concerned about communicating this change correctly, wanting to avoid mismessaging about how much real temperature increase had occurred versus how much was only an apparent increase due to a technical change in measurement. Converting temperature measurements from GMST to GSAT moves the world much closer to the 1.5°C threshold. So even though it makes sense from a perspective of accuracy, the messaging coming out of the new numbers is bleaker than even what was published in SR1.5.

Even after WGI and WGIII resolved to make these changes, remnants of the previous approach, where GSAT and GMST were assumed to be equivalent, could be found all over the AR6 report. For example, the draft "burning embers" diagrams for AR6, which use fiery colors of increasing intensity to designate increasing climate risk, used GSAT when discussing model results and GMST when analyzing observations.

Along with the temperature alignment issue, there was also considerable discussion during AR6 about time frames. This conversation is not new to the IPCC or to climate science and policy in general. Policymakers, especially democratically elected politicians, work on a drastically different timeline than the Earth's response to anthropogenic carbon emissions. What is near term for a prime minister does not correspond to near term for ice sheet disintegration. Generally, scientists work with longer timescales to capture natural phenomena in addition to being able to observe patterns in projected climate change. While a model cannot always deliver reliable annual projections, a decadal or century timescale gives a clearer perspective on the rate and pace of change. To clarify the time scales being used, at WGI LAM2 in January 2019 in Vancouver, a breakout group on reference periods and global mean temperature definitions suggested aligning the following four reference periods with these dates: near term (2021–2040), mid term (2041–2060), long term (2081–2100), and beyond 2100.

The debate over these reference periods continued throughout the AR6 cycle, particularly as experts began to consider how politicians might regard the two-decade timescales, potentially decreasing the sense of urgency around climate action and forestalling ambitious political changes. If 2039 is near term, why would a politician opt to bear the costs now? Conversely, WGIII had another, more policy-relevant definition of near term (to 2030), mid term (up to 2050), and long term (to 2100).

WGI and WGIII spent a lot of time during the pandemic summer of 2020 working out some of the measurement, timeline, and scenario details in a series of meetings and webinars. One area of focus was what ultimately became WGI Figure 1.25 (see our Figure 1).

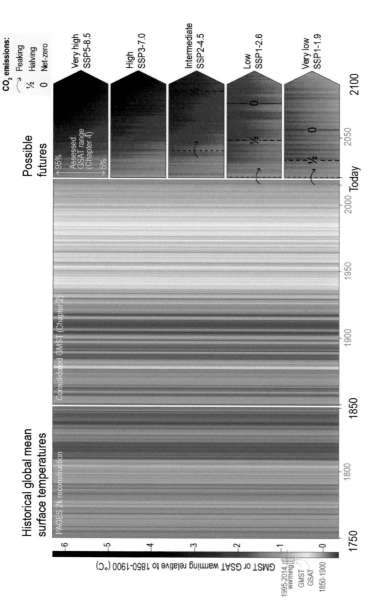

Figure 1 The "warming stripes" of Figure 1.25 in IPCC, 2021: Chapter 1. In: *Climate Change 2021: The Physical Science Basis. Contribution of Working Group I to the Sixth Assessment Report of the Intergovernmental Panel on Climate Change* [Chen, D., M. Rojas, B. H. Samset, et al., 2021: Framing, Context, and Methods. In: *Climate Change 2021: The Physical Science Basis. Contribution of Working Group I to the Sixth Assessment Report of the Intergovernmental Panel on Climate Change* [Masson-Delmotte, V., P. Zhai, A. Pirani, et al. (eds.)]. Cambridge University Press, Cambridge and New York, NY, pp. 147–286, https://doi.org/10.1017/9781009157896.003.]

WGI Figure 1.25 (reproduced as Figure 1 here) was one of the efforts to project global temperature under different scenarios of future emissions, here using Ed Hawkins' global warming stripes. His web interface could be used to generate global warming stripes for particular locations, and they became popular among the climate set post-Paris. For example, knitters began incorporating the stripes into scarves, making wearable art that told the history of anthropogenic warming. In Figure 1, the stripes are used to visualize how the pattern progresses under different scenarios, clearly showing a warming world no matter what choices we make now, but with variation from somewhat warmer to much hotter depending on those choices.

The choice by AR6 authors to shift to GSAT meant that the world appeared a little hotter in the report than it would have if they had stuck with a mixture of surface temperature expressions. This decision – an attempt to be clearer and more accurate – also has political implications as 1.5°C edged a little closer, not only due to carbon emissions but through changes in calculations. At the UNFCCC, Parties push to "keep 1.5 alive," a slogan that urges ambitious climate action and attention to justice: to surpass the 1.5°C threshold signals defeat, loss, and death.

Designing the Carbon Budget

Figuring out how to measure temperature thresholds is challenging in and of itself; and IPCC assessments have recently added another fraught and imprecise indicator, the carbon budget – an estimate of how much greenhouse gas (GHG) humanity can emit before hitting a temperature threshold, be it 1.5°C, 1.7°C, or 2°C. The IPCC began featuring the notion of the carbon budget during the AR5 cycle (Lahn, 2020a), but it was first popularized through the UN Environment Emissions Gap Report. This report, which maps the remaining carbon budget for a given threshold, was first published in 2010 and is updated annually, to great fanfare, at a well-attended side event at each COP. As a schematic, the carbon budget allows scientists creative space to imagine actionable, policy-relevant outcomes (Lahn, 2020b), even if those imaginings fail to become political reality (Geden, 2018). IPCC experts seem to have noted the increasing attention and apparently intuitive policy-relevance of the gap report and sought to take up this narrative within their assessment as well.

The IPCC expanded the carbon budget between AR5 and SR1.5, and slightly again in AR6 WGI, giving the impression of more space for GHG emissions before crossing the 1.5°C threshold. This was highly controversial at the plenary meeting for the WGI Summary for Policymakers (SPM), forcing the creation of a contact group to work on these sentences, in which delegates move outside of the plenary meeting for focused textual revisions. The Earth Negotiation Bulletin reported on the work around the carbon budget in SR1.5, writing:

> On Friday afternoon, contact group rapporteur WG I Vice-Chair Jan Fuglestvedt reported that the paragraph had been revised to improve consistency and to convey uncertainties, and the lower budget figures in the SPM compared to those in AR5. A lengthy discussion ensued as Egypt requested further reference to those numbers in the AR5 carbon budget to allow for consistency, and to avoid the impression that the latest estimations invalidate the carbon budget presented in AR5. The authors explained the methodology used in SR1.5 and how the budget differs from previous estimations, including that it considers direct observations, and non-CO_2 emissions, and is specific to 1.5°C. To explain this difference, various formulations for footnotes were proposed, with participants eventually agreeing to refer to the carbon budget in the SPM and to leave out references to estimations for earlier periods. (2018, p. 9)

In a *Carbon Brief* explainer, Zeke Hausfather (2018) noted that the mismatch between modeled and observed past emissions forced the revision, based on work originally published by a team of IPCC insiders (Millar et al., 2017). Along with the changes in measuring surface temperature, the revisions to calculating the carbon budget are more scientifically accurate and standardized, authors claim, though the changes can also cast doubt on the findings among less expert readers, or those seeking to discredit the IPCC.

Scenarios and Pathways as Possible Climate Futures

Deciding whether the planet has exceeded the 1.5°C threshold or estimating when this may happen is as much a question of measuring the present as it is an exercise in forecasting the future. In the IPCC, this type of work is carried out through scenarios and pathways. These objects are among the most complex and controversial in the whole work of the organization. Scenarios and pathways in AR6 are used to estimate when the planet will exceed the 1.5°C threshold, as well as to assess "to explore future emissions, climate change, related impacts and risks, and possible mitigation and adaptation strategies" (IPCC, 2023a, p. 9).

The adage that a picture tells a story of a thousand words is certainly true of WGI's AR6 Figure 1.27 (2021) (see our Figure 2). Its relatively simple design belies the complexity of how multiple scientific and social scientific disciplines are woven together across WGs. It shows how multiple scenarios or pathways are assessed in IPCC reports, building on the work by different modeling communities. In the center of the figure, we see the global climate models (GCMs) that simulate the Earth's system and are the domain of WGI. These models show how the Earth's atmospheric, oceanic, cryospheric, and terrestrial

systems respond to political and economic decisions and resulting GHG emissions. Working Groups I and II also use models to assess regional climate
changes, impacts, and risks. These models are different from the GCMs used
by WGI and different again from those used in WGIII to explore possible
mitigation pathways; they combine simplified GCMs with economic models
and are called Integrated Assessment Models (IAMs). To produce a coherent
narrative that includes input from all these different types of models, WGI and
WGIII agreed at the outset of AR6 upon a key set of scenarios, called Shared
Socioeconomic Pathways (SSPs), covering "the range of possible future development of anthropogenic drivers of climate change found in the literature"
(SYR SPM, 2023, p. 9). In Figure 2, these can be seen entering the flow in the
top left corner.

These scenarios were crucial when AR6 authors sought to update SR1.5's
assessment that the 1.5°C threshold would likely be reached between 2030
and 2052. Working Group I concluded that in all scenarios considered
(except SSP5-8.5), the median estimate for crossing 1.5°C lay in the early
2030s – ten years earlier than estimated in SR1.5 – primarily due to
methodological and analytical advances. When it came to reflecting those
findings in the SPM, some countries noted in the reviews that they deemed
the message "significant" (Canada) or "sensitive and impactful" (Japan).
Others opposed this. China called the comparison between SR1.5 and AR6
"not sufficiently justified and not convincing enough" and suggested delet-
ing it. Saudi Arabia asked "to omit this finding." The conclusion eventually
disappeared from the text of the summary – though some information related
to the time frame remains in a table.

The combined modeling effort of WGI and WGIII in the Synthesis Report
sought to bring back the question of the time frame by which 1.5°C would be
exceeded. The final draft of the report notes that "continued greenhouse gas
emissions will lead to increasing global warming in the near term, with best
estimates of reaching 1.5°C in 2030–2035 in nearly all scenarios and pathways." But again, some governments opposed the conclusion in the approval
plenary. As a compromise, the time frame was relegated to a footnote, leaving
only the reference to "in the near term" in the main text, which – building on
WGI's definition – means "2021–2040."

Scenarios and pathways were controversial in AR6. Given that they bundle
together possible futures from a range of assumptions about socioeconomic
conditions, they make easy targets for criticism. An Indian delegate noted
during the approval plenary of WGI that the SSPs sounded like policy-
prescriptive language on mitigation (Earth Negotiations Bulletin, 2021). In
response, the authors added to the SPM that "IPCC is neutral with regard to

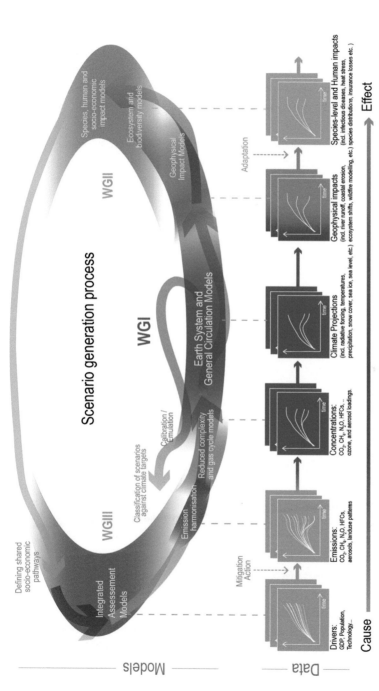

Figure 2 WGI AR6 Figure 1.27 depicting the IPCC scenario generation process. Figure 1.27 in IPCC, 2021: Chapter 1. In: *Climate Change 2021: The Physical Science Basis. Contribution of Working Group I to the Sixth Assessment Report of the Intergovernmental Panel on Climate Change* [Chen, D., M. Rojas, B.H. Samset, et al., 2021: Framing, Context, and Methods. In *Climate Change 2021: The Physical Science Basis. Contribution of Working Group I to the Sixth Assessment Report of the Intergovernmental Panel on Climate Change* [Masson-Delmotte, V., P. Zhai, A. Pirani, et al. (eds.)]. Cambridge University Press, Cambridge and New York, NY, pp. 147–286, https://doi.org/10.1017/9781009157896.003 .]

the assumptions underlying the SSPs . . . ." In a repetitive set of interventions during the approval plenary of WGIII, the same delegate demanded that all references to scenarios and pathways be qualified as "global" and "modeled" pathways to underscore their social construction. Similarly, Saudi Arabian delegates insisted on using the verb "projects" instead of "will" when talking about modeled futures, to emphasize that these are simulations of reality, not reality itself. Pointing to the difference between a socially constructed scenario and the physical or social world it represents is not without basis. Doing so can also raise doubt about the science. However, given that models are inherently socially constructed, effort should be put into being as clear and transparent as possible regarding what they can say about climate change, as well as their limitations.

Scenario building is essential, expert work that helps make comparisons among big model projects possible. Climate scenarios and pathways have proliferated in number and diversity to help decision-makers consider the outcomes of their actions. These scenarios shape present and future storylines: They delimit alternative options and can reveal the cultural and political biases of their creators (Beck and Oomen, 2021). By claiming to be neutral with regard to the assumptions underlying the scenarios and pathways, the IPCC deflects important questions around the futures it prioritizes.

Conclusions

In this section, we showed why simple lines cannot be drawn between the hard facts of science on one side and the contingencies of social processes of assessing science on the other side. We discussed the scientific and political origins of the temperature thresholds, as well as the work in AR6 to measure global temperature, design a carbon budget, and configure agreement between models used to assess physical and economic aspects of climate change. All these complex processes involve multiple contingencies that are implicated in determining a temperature target.

The model-based projections discussed above were complemented by other modes of quantitative analysis that allowed AR6 authors to provide more robust projections than in previous reports. However, with the strengthened science, we more clearly understood the warming that our activities had already committed us. That is, the world is already almost at the temperature threshold agreed in Paris.

The AR6 reports suggested that we were at the edge of 1.5°C but stopped short of stating we had crossed over. The 2022 WGIII SPM notes that to avoid reaching 1.5°C with no overshoot, emissions must peak before 2025. That threshold is next to impossible to avoid crossing – at the time the SPM was published, we were only 1,000 days away from 2025.

Experts outside and adjacent to the IPCC are also concerned with how close we are to the 1.5°C threshold. In 2020, the World Meteorological Organization (WMO) published its "Global Annual to Decadal Climate Update." One note-worthy prediction claimed there was a 20 percent chance that the global average temperature would cross 1.5°C above preindustrial temperature for at least one year over the next five years (Hermanson et al., 2022). Richard Betts, head of climate impacts research at the Met Office Hadley Center, a professor at the University of Exeter, and an IPCC author, wrote a blog post contextualizing this information. Writing about the projections, he said: "While this is another reminder of the Earth's rising temperatures, it is important to note that it would not mean that the Paris Agreement's long-term goal to limit warming to 1.5C will have been breached" (2020). The primary reason for this was that the Paris Agreement limit referred to anthropogenic warming only, while the WMO projections took natural climate fluctuations into account. That is, there is a solid chance that the planet will cross the 1.5°C threshold, edged over by some regular, annual variability, but that does not mean that we will have immediately crossed the path of no return – instead, we will bobble over and under 1.5° of warming from year to year before staying there for a long time.

Even below 1.5°C of anthropogenic warming, our world experiences impacts – displaced people, diminished or changed livelihoods, and the death of tropical coral reefs, to name some examples. But as the IPCC says, every fraction of a degree matters – and fighting to limit warming as much as possible and as quickly as possible averts most catastrophic climate risks. In other words, 1.5°C is just a number. It should not detract from the necessity to work toward the social, economic, and political changes necessary to reduce anthropogenic climate change.

4 Building Bridges between the Physical and Social Sciences of Climate Change

When the Intergovernmental Panel on Climate Change (IPCC) was established, one of its primary objectives was to discern whether humans were having an impact on the Earth's climate. Since then, the IPCC has found that humans are indeed changing the climate, and after the 2015 Paris Agreement, there is now much greater emphasis on formulating response options. To this end, one of the significant changes that the IPCC made for its Sixth Assessment Report (AR6) is to place greater emphasis on integrating knowledge between Working Groups

(WGs). This is not a simple task; the IPCC was formed with an "epistemological hierarchy" that privileged the physical sciences (O'Neill et al., 2010). Working Group I assesses the physical sciences of climate change. Working Group II assesses climate impacts, human vulnerability, and adaptation to climate change from perspectives of both the natural and the physical sciences. The two WGs had different ways of envisioning themselves and their tasks, and prior to AR6, there was little collaboration between them. But the nonlinear quality of climate change, which includes the capacity for risks to cascade through several inter-connected social-ecological systems simultaneously, means that surprises and unexpected climatic events are becoming more likely. This, in turn, makes it important to understand how people's vulnerability and capacity for adaptation are prefigured by social, cultural, political, and economic structures, which are dealt with in WGII. This desire for the natural and social sciences to work more closely to respond to the changing quality of risk is one of the reasons why the IPCC Bureau prioritized integration between WGs in AR6.

The authors featured in this section made concerted efforts to reform rela-tionships and put the physical and social sciences on a more equal footing. As we show, assessing scientific knowledge cannot be reduced to a simple algo-rithmic procedure. Rather, it requires the human capacity to collaborate, com-municate, and find creative solutions to complex problems. In this section, we see WGI and WGII authors doing just that as they build bridges between their WGs.

This matter came to our attention during the first Lead Author Meeting (LAM) for WGI, convened in June 2018 in Guangzhou, China. A co-chair addressed the assembled authors gathered in the large conference hall. Standing at the podium in front of close to 200 attendees who were sitting at rows of tables, with two massive screens livestreaming the presentation on either side of them, they spoke about the need to integrate WGI more consistently with other WGs throughout the AR6 cycle. This might have surprised many in the room because there has been a clear separation between WGs in the IPCC from its inception. Each WG is coordinated by a different Technical Support Unit (TSU), and these are headquartered in, and funded by, different nations. Each WG has its own meetings and each is dominated by different academic discip-lines. The "epistemological hierarchy" (O'Neill et al., 2010) informs the linear model of science for policy, as well as the timing of how each WG operates within the assessment cycle, with WGI beginning their work first and WGIII beginning their work last. Historically, it was rare for IPCC authors to have much involvement with WGs outside of their own; efforts to coordinate across WGs usually fell to the Secretariat or task forces (Vardy, 2022). But at LAM1, the co-chair was upsetting this entrenched way of doing things by insisting that rather than maintain this separation of WGs, it was the duty and obligation of IPCC authors to try to bridge it.

The co-chair's announcement didn't come out of the blue. A lot of work over the previous two years had already gone into setting the stage for WG integration. Specifically, the AR6 chapter outlines, which were approved by the IPCC plenary in September 2017, already contained a structure of integration. And, the effort of integrating knowledge across WGs had already been put to the test and found by many to be successful in the Special Report on Global Warming of 1.5°C (SR1.5; IPCC, 2018). Indeed, SR1.5 demonstrated, to many of its authors, the benefits of collaborating across scientific disciplines. For example, we spoke with a political scientist who was a Lead Author (LA) in SR1.5 and a Coordinating Lead Author (CLA) in AR6. She told us how the historical structure of the IPCC "creates its own inertia, its own assumed worldview, its siloed work around the three Working Groups." But she went on to say that the experience of working on the SR1.5 disrupted these siloed patterns and was "liberating [because . . .] everybody was mixed up together, problem-solving." This sentiment – that involving authors from all three WGs in the chapter writing teams for the SR1.5 was beneficial – was repeated by many of our other interlocutors.

While a handful of WGI AR6 authors had experienced the benefits of integration through their involvement in SR1.5, many more had not. The challenge faced by the co-chair when they addressed WGI authors in Guangzhou, then, was to convince them of the benefits of integration. This was not a trivial challenge; instead of working directly with authors from other WGs in their chapters, as was done in the SR1.5, they were asking authors in WGI to consider integrating with authors who they didn't necessarily know, who they might never meet, and who were working on timelines and in places removed from their own. In effect, the structure of siloed WGs remained in place, but they were asking for authors to achieve integration, nonetheless.

Bridging Cultures of Separation

Several of the people involved in the scoping meeting for AR6 became self-described ambassadors in the writing processes that unfolded between 2018 and 2021. As we attended the LAMs, we noticed a few of these ambassadors showing up to attend the LAMs of WGs other than their own. They would frequently be introduced in the plenary sessions, and reference would be made to their mission of trying to integrate knowledge.

We conducted interviews with these ambassadors, including a WGI Vice-Chair who was involved in the planning for AR6. He spoke about how his experience in AR5 led him and others to the conclusion that they should begin to collaborate more deliberately across WGs as early in the assessment writing process as possible. He also spoke about the process of championing a structure that embeds integration across WGs in chapter outlines in AR6. Instead of chapters each focusing solely on their own individual content, as they had tended to do in previous assessments, the structure of AR6 asked authors to make connections with chapters and sometimes WGs other than their own. But within that structure, the writing of the actual assessment report is still up to the authors in each chapter. As the WGI Vice-Chair told us:

> It's up to the author team to take the outline further. The outline that was approved has indicative bullets, as we call it, and underlying guidance text. So it's up to the author team. And they can use that freedom. And they can choose not to use it. They can still produce a good report [if they do not use the guidance text], but the opportunities are larger [if they do].

In other words, if authors are not persuaded of the viability or the vision of integration presented in the chapter outline, they cannot be forced to follow it.

A significant aspect of integrating knowledge is overcoming differences between WGs. For example, we spoke with a statistician and climate modeler from WGI who was attending a WGII LAM as an ambassador. She said she was aware of the IPCC's history of having entrenched "cultural differences" between WGs. Asked to elaborate, she said:

> I feel there is, in a lot of older generation scientists, there is this idea that everything that has to do with the human dimension is somehow, to use another technical word, wishy-washy. That it's not science. It's something that is more arbitrary and it's fuzzy, especially the non-quantitative side of the human dimension work on global change. [It] does not rise to the same level of physical science.

She stressed that this was not her view, but that of others with whom she worked in AR5. This was echoed by another WGI author, a climate physicist, who spoke to us at a WGII LAM. He too had been involved in AR5, and said, in a rather understated way: "I think in the Fifth Assessment that we took more of an isolationist approach."

This history of separation between WGs was confronted in AR6. Another ambassador, a mathematician and climate modeler from WGI, told us that he felt he had been regarded with suspicion by WGII authors when he first showed up at their LAM. That is, some WGII authors might have felt that he was going to foist a particular approach on them. He said he tried to deflate the tension and then continued:

But coming back here [WGII LAM2] for some reason, there's a level of familiarity somehow because it's our second time. And that means therefore that some relationship has been established like, "Oh yeah, you're right. Yeah." And so therefore that means that the interaction that we had then helps again [with] the interaction that we're having now. And there's a level of familiarity and maybe a level of confidence as well in dealing with each other for some of the people that we've connected with in Durban [at LAM1]. And ... the connection made then carries through [to] now. And then it just makes the process more optimistic in terms of moving forward.

This effort was noticed by some WGII authors. For example, a specialist in urban vulnerability from WGII told us:

[I am impressed by the] enthusiasm and openness about working across the Working Groups, both here and in Durban [at LAM1]. The WGI folks just getting stuck in [and] being very incredibly willing to engage without doing anything that privileges the sort of science that they do in comparison to the sort of science that we do.

A WGII political scientist also told us how the effort to bridge cultures of separation is different from the norms of working as an individual researcher in her own laboratory, where she has an intimate knowledge of the research problem. Working in the IPCC, she said, meant that she had to work outside of her own comfort zone, to learn about new concepts and approaches in other WGs. This brings up the importance of building relationships: "Increasingly, I'm finding – in most of my other work as well as this – that about 70 percent of international research collaboration is building human relationships and trusting working relationships, and 30 percent is the actual research."

The importance of building relationships should not be underestimated. A WGII earth systems scientist told us how some people who volunteered for previous IPCC assessments had bad experiences with experiments in integration. But, he continued, AR6 brought in more first-time authors with no prior experience in the IPCC than any other assessment in the past. That meant there were more people willing, in AR6, to make a "handshake" (share information and strategies for communicating that information) than in previous assessments. This conversation took place at the WGI LAM2 in Vancouver, before WGII had met for the first time. Although he was a WGII author, he wanted to start coordinating with the natural scientists in WGI so that they could frame the physical science of risk in terms that would align with the risk framework in WGII. He was involved in the scoping meeting, but he said that he knew integration would only succeed if there were enough people in both WGI and WGII who were oriented to a common vision and committed to actively working toward it. He told us that part of his challenge was to ensure the risk framework was embedded organizationally in both WGI and WGII:

> This [WGI LAM] is going really well, and the next thing is like, okay, how do we—so, how's this gonna go [at the WGII LAM]? And like, how to make these links? I don't want to chase individual people around. Do we want to do it in an institutional way and get [it] on the agenda?

In referencing the "institutional way" and getting the risk framework "on the agenda," he was speaking about the desire to structure the workflow across WGI and WGII so that risk would be understood similarly by authors no matter their WG. To this end, building trust and establishing a common vocabulary that could aid shared understanding was crucial. As a WGI climate physicist explained:

> We don't have quite the same language [as authors in other WGs] so that can also be harder, too. And—and trying to collaborate in unfamiliar ways takes more time and [is] even harder and you have to spend more and more time getting to speak each other's language. . . . in these collaborations, you have to sit down together and talk a long time before you get to know them.

One of the vocabularies through which integration was worked toward between WGs is that of risk.

Constructing a Framework for Risk

The importance of paying attention to how the IPCC integrates WGI and WGII becomes clear if we consider the example of risks related to changes in the Earth's hydrological cycle. The monsoons on which most smallholder farmers in the Global South rely are treated from a biophysical point of view in WGI. But WGII is where you will find the regional chapters that focus on impacts, adaptation, and vulnerability in parts of the world where the changes to monsoons impact the lives of people who rely on them for their livelihoods. The separation of the physical properties of monsoons and the livelihoods of smallholder farmers could maintain the fiction of an "ontological dualism," that is, that the social and natural worlds really are separate entities (Pickering, 2009). From this perspective, climate change exists in nature (covered in WGI) and impacts exist in society (covered in WGII). But when they are integrated through efforts such as those seen in AR6, difficult questions arise. For example, should authors prioritize the social and political conditions such as poverty, gender, and racial discrimination that prefigure vulnerability to climate events, or the climate events themselves? Put simply, does risk emanate from the parts per million of carbon dioxide in the atmosphere – and physical processes related to climate change – or does risk come from the inequalities created by political, economic, and social forces? And what makes smallholder farmers in the Global South resilient to disruptions in predictable and regular patterns of precipitation? If we maintain the ontological dualism that locates the problem of climate change in the natural world, the only way to enhance the resilience of the smallholder farmers is to reduce global greenhouse gas (GHG) emissions. But considering the natural and social worlds as an integrated whole allows us to focus not only on mitigation but also on the political, economic, and cultural structures that situate the farmers as vulnerable in the first place.

Sitting in the restaurant of a Durban hotel, a WGII specialist in urban risk reduction spoke to us about decades of research he had done with urban poor in Caribbean nations and Sub-Saharan Africa. That work helped shape the issues that he thought should receive attention in AR6. He talked about how the urban poor in Sub-Saharan Africa are often already so marginalized that they are dangerously susceptible to even the ordinary "natural variability" in the climate, never mind the increased frequency and intensity of climatic events associated with anthropogenic change. For example, he said, even a minor flood can inundate neighborhoods with standing floodwater contaminated by sewage:

> so in Sub-Saharan Africa . . . a lot of people suffer, and certainly their health is compromised by chronic exposure to environmental conditions, so a little bit of rain if you're very, very vulnerable will lead to some flooding, and that will affect your health.

It is well accepted that climate change can produce risk through physical hazards. But what about the risk that is produced through social, economic, and political issues such as poverty? How should the IPCC deal with the forces that lead people to living with "chronic exposure to environmental conditions . . . that will affect your health"?

This was not the first time such ideas had been brought up. Indeed, the United Nations (UN)'s Sustainable Development Goals (SDGs) were included in the scoping of AR6. The author said:

> This is an opportunity basically to meet the SDGs through climate change adaptation. . . . climate change is an expression of failed development. It's not a standalone agenda. And I know there's a lot of pushback from that, from climate scientists who really want to see a very clear-cut space for climate science.

WGII assesses peer-reviewed publications that include work in political ecology and critical global development studies. From this perspective, scholars argue that the extraction of surplus value from the Global South should be included in the consideration of risk. Speaking in broad terms, some WGII authors argue that the dynamics of global development should be analyzed alongside the dynamics of climatic change. They might contend, for example, that they should assess the globalized financial flow of debt that entraps smallholder farmers in the Global South who can no longer rely on precipitation patterns and are driven to borrowing money to drill ever-deeper wells for irrigation that depletes groundwater aquifers. Stuck in the cycle of expanding debt, such smallholder farmers can be forced off their land – more by the political economy of debt than by climate change (Taylor, 2014). As the WGII specialist in urban risk reduction put it:

> The impact of failed development [is more severe] by orders of magnitude, as opposed to environmental change. So it's a chronic condition. It's not discrete [climatic] events. ... So if one is focusing on literature around discrete [climatic] events you miss the risk facing the urban poor in Africa.

These issues are at the heart of long-standing academic debates about the environmental determinism of framing climate change exclusively as a biophysical problem. O'Brien et al. (2007) argue that while it is true that climate change can result in "outcome vulnerability" through direct impacts on people, the social, economic, political, and historical circumstances in which people are embedded, or what they call "contextual vulnerability," has a profound influence on people's ability to respond to climate stresses. These concepts make it clear that sometimes the best way to fight climate change is to address how people are situated within the globalized political economy.

The consequences of how concepts such as "social transformation" and "transformative change" are used by the IPCC are not trivial. The IPCC has an influence beyond providing science advice for decision-makers, including influencing the agenda of international development agencies, along with the broader philanthropic and corporate arms of the development industry. Therefore, IPCC authors are not simply trying to assert disciplinary dominance to satisfy their own egos; rather, they are trying to reframe the narratives through which climate change is discussed to foster social and political responses. They are reflexively trying to change the narrative through which the IPCC frames climate change.

The difficulties of coordination, building trust, and reframing narratives are influenced by the IPCC's operating procedures. For example, the IPCC is compelled by its mandate to provide "traceability" or a citation pathway that shows where its information comes from. Thus, authors often find themselves constrained by the disciplinary norms, standards, frameworks, and methods – the grammar and syntax, if you will – of the particular discipline that they are assessing. If IPCC authors want to integrate across WGs, they are faced with the daunting task of establishing how sometimes incommensurate literature can be compared fairly, which authors described to us above as taking place through the social processes of gaining trust with others and figuring out how to communicate. Here is how one of the ambassadors from WGI quoted above, the mathematician and climate modeler, put it: "From a climate information perspective, there is still a gap . . . but it's more of a communication gap than an actual information gap, and how you integrate the different knowledge types. . . . it's the integration of all of the elements of risk." But it is not a simple or straightforward task because, as indicated above, the integration of "all the elements of risk" could include the political economy of "failed development."

Given the political stakes of framing risk, it is little wonder that it took a lot of time and effort to integrate across WGs. We spoke with a physicist in WGI about the task of compiling a common risk framework across WGI and WGII. She said that it took a lot of work, with hundreds of emails over many months, to stabilize a definition of the risk framework that everyone seemed happy with. And then someone would put forward an amendment or challenge to the risk framework that would prompt another long round of debate and deliberation. She said that it was only "five days before the first-order draft had to be handed in" that they achieved a "consensus definition" on what the risk framework consisted of.

Conclusions

As shown above, it took a considerable amount of work to integrate knowledge between WGs that were meeting at different times and places from one another. But what difference did this make for how risk is discussed in AR6? What was the payoff? The short answer is this: AR6 developed the risk framework so that it now includes human responses to climate change. While AR5 assessed risk in terms of physical hazards, exposure, and vulnerability, AR6 built upon this to establish a new risk framework (Simpson et al., 2021). As stated in the Summary for Policymakers (SPM) for WGII, "the risk that can be introduced by human responses to climate change is a new aspect considered in the risk concept" (IPCC, 2022a, p. 5). The style and tone in which the thirty-four-page WGII SPM is written mask some of the more interesting points relevant to including human responses in the risk framework. But one of the differences made by including human responses to climate change as an additional risk factor is that it brings maladaptation into our purview. That is, it allows for the analysis of how social, political, policy, and economic responses to climate change can exacerbate vulnerability and contribute to further climate impacts (Schipper, 2020).

While including human responses in the risk framework is a noteworthy outcome of integration between WGs, another significant outcome could be a long-term change in the culture of the IPCC. That is, it might have seeded a trend toward further integration. We spoke with a co-chair about how the shift toward integration in the IPCC mirrors changes in the wider academic world. She said:

> the community has changed. Not everyone, but it has. And the borders are not that clear. If you ask someone "Are you a Working Group I or Working Group II scientist?" sometimes they don't know. . . . . That's how it is in the research community.

This statement demonstrates the tension between how science works and the way that the IPCC is organized institutionally. While it may have made sense in the late 1980s and early 1990s to create the IPCC with the structure of separate WGs, it makes less sense now for the scientists whose work increasingly crosses disciplinary divides.

The IPCC's inertia makes large-scale organizational changes difficult; based on the interviews that we conducted, it seems likely that the structure, in terms of having three WGs will remain in place. In other words, the organizational and institutional framework of the IPCC will continue to bear the imprint of ontological dualism. But set against this are compelling examples of what can be achieved through integration. In this section, we showed how the imperative to integrate knowledge across WGs played out in the case of the risk framework. The resulting addition of "human responses" to the risk framework marks a significant development. To achieve it required authors to engage with heavily weighted issues of how climate change should be framed in the first place. It also required authors – to greater or lesser degrees – to leave their comfort zones, learn to speak each other's language, and trust one another.

5 How Does the IPCC Define Climate Solutions?

Human societies are so tightly interwoven with the forces that produce anthropogenic climate change that solutions to the climate problem are far from simple. Indeed, all of the solutions that the Intergovernmental Panel on Climate Change (IPCC) assesses include not only scientific and technological but also societal choices, with all the debates associated with decision-making, including economic and political risks and gains. As time goes on, the problem of climate change gets worse, which raises the stakes and makes it all the more important to create lasting changes that will solve it. And, as we discuss below, the continual addition of greenhouse gasses (GHGs) to the Earth's atmosphere means that some of the solutions that the IPCC assesses become less feasible over time.

In this section, we discuss some of the solutions that can be found in IPCC reports. Climate solutions can be found in adaptation, responding to changes that have already occurred, or mitigation, preventing further climate change. Adaptation, traditionally the domain of Working Group (WG) II, is defined as a set of solutions in human systems that address existing and projected climate impacts – or, as the WGII glossary puts it, "the process of adjustment to actual or expected climate and its effects, in order to moderate harm or exploit beneficial opportunities" (IPCC, 2022a, p. 2898). Mitigation of climate change, primarily the domain of WGIII, is defined as "a human intervention to reduce emissions or enhance the sinks of greenhouse gasses" (IPCC, 2022b, p. 1808). This translates to lowering amounts of GHGs through technological and behavioral change as well as removing GHGs already in the atmosphere. While we do make some reference to discussions that took place in WGII on adaptation options, this section mainly focuses on mitigation options.

With its Sixth Assessment Report (AR6), the IPCC tried to strengthen the conversation around solutions. The IPCC Chair Hoesung Lee was elected in part due to his vigorous focus on solutions, which he carried out up to and through the Synthesis Report. The IPCC had seen a variety of solutions before – they form the heart of WGIII and are also assessed in WGII – but the championing of this piece of climate knowledge across the organization has been new. It reflects the development of climate science: We know, with broad consensus, that climate change is happening, and why. The increased focus on solutions also reflects the realities of climate action: Globally, governments are putting the Paris Agreement into action; and institutions, governments, organizations, corporations, communities, and individuals are – to varying degrees and in various ways – trying to adapt to the toll of climate change and move forward in our warming world.

This strong support for solutions from the top down has shifted the power of the IPCC more firmly away from the traditional center, in WGI, toward WGIII. The more the IPCC delves into the socioeconomic dimension of climate change, the more it exposes itself to scrutiny. The IPCC's institutional practices require authors to avoid policy prescription, so climate solutions must be assessed carefully to avoid suggestions. However, as the climate crisis has amplified, the science is increasingly clear about what actions have become imperative to solve the problem. Yet just because the IPCC deems some solutions possible in a technical sense doesn't mean that they're socially or politically acceptable.

How we think about climate solutions depends on how we understand the climate problem. Entrenched cultural problems, like over-consumption, require cultural change. An energy grid based on fossil fuels can switch to renewables – solving a technological problem with a technological solution. This section first examines how solutions are conceptualized in IPCC reports, as spaces or pathways. Then, we analyze two case studies. The first involves carbon dioxide removal, a controversial solution that some argue is made inevitable through our inaction up to now; the second pertains to demand-side mitigation – which offers a productive counter-narrative to the supply-side energy economics that has dominated IPCC solutions throughout its history. IPCC authors during AR6 worked to broaden their assessment and communication of solution options: At this point, it is imperative to implement multiple solutions simultaneously to reach, or even approach, Paris Agreement targets.

Solution Spaces and Pathways

The scientific community outside of the IPCC has discussed the "solution space" at length, which they refer to as "the space within which opportunities and constraints determine why, how, when, and who" to pursue climate action (Haasnoot et al., 2020, p. 36). This conversation, which was in full swing during the writing of SR1.5, influenced how the IPCC conceived of solutions, as well as how adaptation and mitigation fit together into that space. In an interview, a co-chair explained how the Special Report, the "first joint effort of all Working Groups," provided an opportunity to tell the full story of options for limiting warming to 1.5°C that would have been impossible if only two of the WGs participated (which is how all previous Special Reports were structured). The co-chair explained,

> All of a sudden there's a full storyline emerging that we can contribute to . . . we can take a society into what we call the solution space. We can address what adaptation can contribute to that solution place, a space. We come away from the 1.5 report already with the knowledge of the need for a huge effort into mitigation emissions reduction. There's no excuse for not doing it. There's no alternative to that. And mitigation and adaptation have to be combined to cover the full scope of what needs to be done because we have climate change in action.

While AR6 did not frequently use the term "solution space," WGII, and in particular its Chapter 18, sought to convey similar ideas. More specifically, it built on discussions in AR5 about climate-resilient development (CRD) as the "process of implementing mitigation and adaptation measures to support sustainable development for all" (IPCC, 2022a, p. 2657). Working Group II introduced Figure SPM.5 (see our Figure 3) to illustrate the "rapidly narrowing window of opportunity to enable climate-resilient development" (30). The figure and its intent have been widely shared and are also featured in the Synthesis Report.

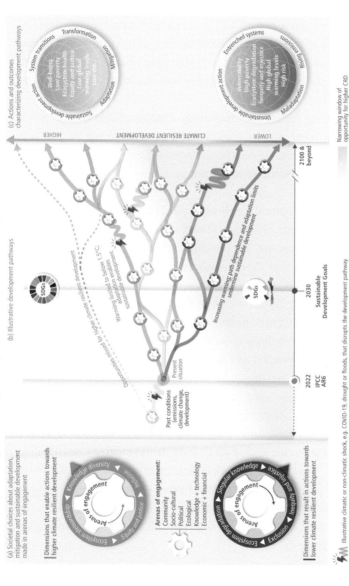

Figure 3 Trajectories for climate resilience development. Figure SPM.5 in IPCC, 2022: Summary for Policymakers [H.-O. Pörtner, D.C. Roberts, E.S. Poloczanska, et al. (eds.)]. In: *Climate Change 2022: Impacts, Adaptation, and Vulnerability*. Contribution of Working Group II to the Sixth Assessment Report of the Intergovernmental Panel on Climate Change [H.-O. Pörtner, D.C. Roberts, M. Tignor, et al. (eds.)]. Cambridge University Press, Cambridge and New York, NY, pp. 3–33, https://doi.org/10.1017/9781009325844.001.

The figure is less about listing and assessing a range of solutions than about mapping the forces and strategies that enable or constrain their deployment and efficacy over time and across contexts. The caption describes how climate-resilient development pathways are the result of cumulative societal choices and actions within multiple arenas (IPCC, 2022a, p. 31). Panel (a) notes the dimensions that result in actions toward higher or lower climate-resilient development. Panel (b) describes the cumulative societal choices that shift global development pathways toward higher or lower climate-resilient development. Finally, panel (c) lists the actions and outcomes that characterize higher or lower climate-resilient development.

The strong framing around the "narrowing window of opportunity" sends a clear signal that some pathways and options soon might not be available anymore. The dashed green line even suggests that some development pathways toward higher climate-resilient development have been eliminated because of past choices.

Climate-resilient development is only one type of pathway used in the IPCC to communicate about climate solutions – though an important one that speaks to Article 2 of the Paris Agreement. The proliferation of qualitative and quantitative pathways and scenarios that depict outcomes of sets of policy decisions is dazzling. These include mitigation scenarios, non-overshoot pathways, overshoot pathways, adaptation pathways, development pathways, sustainable development pathways, shifting development pathways, shifting development pathways to sustainability, illustrative mitigation pathways, and transformation pathways. Embedded into these pathways is a multitude of possible futures, stepping stones of decisions to make now and the worlds they can make possible.

Pathways as an epistemic and assessment tool to describe key characteristics of possible climate policy futures became ubiquitous during the AR5 cycle and rapidly matured as a means for communicating climate solutions in SR1.5. There, the authors' attention shifted away from Representative Concentration Pathways (RCPs), which were numerical proxies of several tiers of climate policy decisions that resulted in estimates of future greenhouse gas emissions and resulting temperature and other changes. In SR1.5, rather than RCPs, the figure that experts most consistently displayed in their presentations was Figure SPM3b (see our Figure 4).

This figure shows different ways – building on Integrated Assessment Models (IAMs) – to hold warming to 1.5°C, each containing not only differing mixes of solutions but entire sets of values and assumptions about the environment, technology, and the economy. The four illustrative model pathways exemplify clear climate communication. Anyone can look at them and come away with a sense that the four pathways involve serious commitments to different bundles of solutions.

Characteristics of four illustrative model pathways

Different mitigation strategies can achieve the net emissions reductions that would be required to follow a pathway that limits global warming to 1.5°C with no or limited overshoot. All pathways use Carbon Dioxide Removal (CDR), but the amount varies across pathways, as do the relative contributions of Bioenergy with Carbon Capture and Storage (BECCS) and removals in the Agriculture, Forestry and Other Land Use (AFOLU) sector. This has implications for emissions and several other pathway characteristics.

Breakdown of contributions to global net CO₂ emissions in four illustrative model pathways

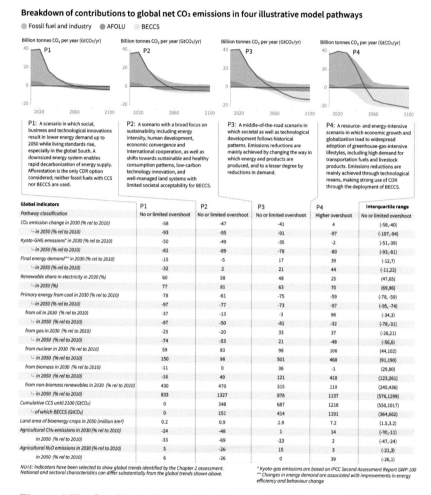

P1: A scenario in which social, business and technological innovations result in lower energy demand up to 2050 while living standards rise, especially in the global South. A downsized energy system enables rapid decarbonization of energy supply. Afforestation is the only CDR option considered; neither fossil fuels with CCS nor BECCS are used.

P2: A scenario with a broad focus on sustainability including energy intensity, human development, economic convergence and international cooperation, as well as shifts towards sustainable and healthy consumption patterns, low-carbon technology innovation, and well-managed land systems with limited societal acceptability for BECCS.

P3: A middle-of-the-road scenario in which societal as well as technological development follows historical patterns. Emissions reductions are mainly achieved by changing the way in which energy and products are produced, and to a lesser degree by reductions in demand.

P4: A resource- and energy-intensive scenario in which economic growth and globalization lead to widespread adoption of greenhouse-gas-intensive lifestyles, including high demand for transportation fuels and livestock products. Emissions reductions are mainly achieved through technological means, making strong use of CDR through the deployment of BECCS.

Global indicators	P1	P2	P3	P4	Interquartile range
Pathway classification	No or limited overshoot	No or limited overshoot	No or limited overshoot	Higher overshoot	No or limited overshoot
CO₂ emission change in 2030 (% rel to 2010)	-58	-47	-41	4	(-58,-40)
↳ in 2050 (% rel to 2010)	-93	-95	-91	-97	(-107,-94)
Kyoto-GHG emissions* in 2030 (% rel to 2010)	-50	-49	-35	-2	(-51,-39)
↳ in 2050 (% rel to 2010)	-82	-89	-78	-80	(-93,-81)
Final energy demand** in 2030 (% rel to 2010)	-15	-5	17	39	(-12,7)
↳ in 2050 (% rel to 2010)	-32	2	21	44	(-11,22)
Renewable share in electricity in 2030 (%)	60	58	48	25	(47,65)
↳ in 2050 (%)	77	81	63	70	(69,86)
Primary energy from coal in 2030 (% rel to 2010)	-78	-61	-75	-59	(-78,-59)
↳ in 2050 (% rel to 2010)	-97	-77	-73	-97	(-95,-74)
from oil in 2030 (% rel to 2010)	-37	-13	-3	86	(-34,3)
↳ in 2050 (% rel to 2010)	-87	-50	-81	-32	(-78,-31)
from gas in 2030 (% rel to 2010)	-25	-20	33	37	(-26,21)
↳ in 2050 (% rel to 2010)	-74	-53	21	-48	(-56,6)
from nuclear in 2030 (% rel to 2010)	59	83	98	106	(44,102)
↳ in 2050 (% rel to 2010)	150	98	501	468	(91,190)
from biomass in 2030 (% rel to 2010)	-11	0	36	-1	(29,80)
↳ in 2050 (% rel to 2010)	-16	49	121	418	(123,261)
from non-biomass renewables in 2030 (% rel to 2010)	430	470	315	110	(245,436)
↳ in 2050 (% rel to 2010)	833	1327	878	1137	(576,1299)
Cumulative CCS until 2100 (GtCO₂)	0	348	687	1218	(550,1017)
↳ of which BECCS (GtCO₂)	0	151	414	1191	(364,662)
Land area of bioenergy crops in 2050 (million km²)	0.2	0.9	2.8	7.2	(1.5,3.2)
Agricultural CH₄ emissions in 2030 (% rel to 2010)	-24	-48	1	14	(-30,-11)
in 2050 (% rel to 2010)	-33	-69	-23	2	(-47,-24)
Agricultural N₂O emissions in 2030 (% rel to 2010)	5	-26	15	3	(-21,3)
in 2050 (% rel to 2010)	6	-26	0	39	(-26,1)

NOTE: Indicators have been selected to show global trends identified by the Chapter 2 assessment. National and sectoral characteristics can differ substantially from the global trends shown above.

* Kyoto-gas emissions are based on IPCC Second Assessment Report GWP-100
** Changes in energy demand are associated with improvements in energy efficiency and behaviour change

Figure 4 The four illustrative model pathways to limiting warming to below 1.5°C. Figure SPM.3b in IPCC, 2018: Summary for Policymakers. In: *Global Warming of 1.5°C. An IPCC Special Report on the Impacts of Global Warming of 1.5°C above Pre-industrial Levels and Related Global Greenhouse Gas Emission Pathways, in the Context of Strengthening the Global Response to the Threat of Climate change, Sustainable Development, and Efforts to Eradicate Poverty* [Masson-Delmotte, V., P. Zhai, H.-O. Pörtner, et al. (eds.)]. Cambridge University Press, Cambridge and New York, NY, pp. 3–24, https://doi.org/10.1017/9781009157940.001.

The four illustrative model pathways emerged as the dominant message of SR1.5: There are ways to hold warming to 1.5°C and these require attention to and from various sectors. The way we live now – in resource- and energy-intensive societies – is exemplified in the rightmost figure. The brown and yellow curves represent the activities required to compensate for residual emissions and achieve net negative emissions – through removals in the AFOLU (Agriculture, Forestry, and Other Land Use) sector (e.g., afforestation and reforestation) and Bioenergy with Carbon Capture and Storage (BECCS).

Solving the climate problem without transformational economic activity (the third pathway), or complete changes in our dominant cultural systems (the first pathway), requires a technological intervention that most nonexperts have not heard of. While the methods have been around since the 2000s, afforestation/reforestation and BECCS are not presently implemented anywhere near the scale of P2, P3, or P4, are expensive, and could have negative impacts on biodiversity and water and food security.

Pathways are tested against their likelihood not to exceed a certain global warming level, but not against the likelihood of being implemented. Several authors raised serious doubts about what it meant, socioeconomically speaking, to bend the curve as characterized in the pathways. They are "unrealistic. It is a free fall," noted one author in a meeting. Another author told us:

> If you go this [miming the curves of the pathways], that requires a miracle, that requires people to function fundamentally different to how they functioned in other situations. And yeah, your integrated assessment models show that's the cost-effective way to get to where we need to go. But fuck it, you need a political miracle to make that happen, and I don't think you get to have miracles happening. I think you fundamentally have a difference of seeing the world between those who say, "Well, I'm not a political scientist" Versus people like me who say, "Well, I'm sort of a bit of everything and I don't have time to write IAMs, but I can tell you, the curve is going to have to look like that" [miming a less steep curve].

Climate scenarios and pathways have proliferated in number and diversity to help decision-makers consider the outcomes of their actions. They shape present and future storylines: They delimit the "possibility space" (Cointe, 2022, p. 137) and can reveal the cultural and political biases of their creators (Beck and Oomen, 2021).

Case Study: Carbon Dioxide Removal

While often presented as default options in the pathways, BECCS and afforestation/reforestation are among a wider range of "technologies, practices, and approaches that remove and durably store carbon dioxide from the atmosphere," known under the umbrella term of carbon dioxide removal (CDR) (IPCC, 2023).

Carbon dioxide removal has gained steam as an inevitable policy option, and the IPCC has played a key role in normalizing it. Given the rapidly shrinking carbon budget and the risk of overshooting 2°C, the AR5 WGIII introduced CDR methods in emissions scenarios reaching 2°C, and in particular BECCS and afforestation/reforestation. "The story, according to some people, is that [CDR] was invented in AR5 to see how we could stay below the lower [temperature] levels," an author told us. The publication in 2018 of SR1.5 further inscribed CDR in all its scenarios limiting warming to 1.5°C, as described above.

In AR6, WGIII included a chapter on "cross-sectoral perspectives" that covered topics such as food systems, land systems, and CDR. Carbon dioxide removal methods were discussed at length and key conclusions were included in the report's Summary for Policymakers (SPM). Headline statement C.11 in particular states that "the deployment of CDR to counterbalance hard-to-abate residual emissions is unavoidable if net zero CO_2 or GHG emissions are to be achieved" (IPCC, 2022b, p. 47). The paragraph introduces two main arguments. First, it is illusory to think that we can live in a world entirely without CO_2 emissions. There will be "hard-to-abate residual emissions," including those from agriculture, aviation, shipping, and industrial processes. And, second, the removal and durable storage of CO_2 in geological, terrestrial, or ocean reservoirs, or in products, is necessary if governments want to achieve "net zero" CO_2 or GHG emissions – the long-term objective guiding most national climate policies today (Van Coppenolle et al., 2022).

As a headline statement, C.11 ranks high in the hierarchy of SPM statements and conveys important messages that the IPCC wants to transmit to decision-makers, the media, and the public. The shift to "net zero" framing has been notable since the publication of AR5 and reflects a major epistemic shift, including a better understanding of the global temperature response to changing atmospheric GHG concentrations and of the global carbon cycle. The IPCC acted as a knowledge broker, successfully communicating this novel understanding to policymakers.

If net zero has progressively become an "article of faith" (Allen et al., 2022, p. 28), so too has CDR, as the language around its "unavoidable" deployment seems to suggest. We encountered plenty of criticism and resistance about the topic within the IPCC and among the WGs. An IPCC author noted their concern during the first WGII Lead Author Meeting (LAM1), trying to nudge a conversation that already seemed entrenched from the outset:

> Somebody from the Global South, that is a woman, had said we shouldn't be accepting CDR, necessarily. It may not be socially acceptable. And a lot of people had quietly sat there and thought, you know, we don't agree with this [CDR], but the narrative was still dominant.

Another also noted concern about BECCS as a solution during a WGIII LAM. As this author put it:

> There was a great reliance upon BECCS to pull down the emissions . . . for the SR1.5 report, you know, some of the two-degree scenarios were looking at 15 million square kilometers of additional land for bioenergy carbon capture and storage, or seems to be, I don't know why the IPCC would even entertain that. It seems to be ridiculous.

Similar concerns were raised during the SPM approval sessions. For instance, during the approval of the WGII SPM, which assessed risks arising from response measures (including CDR methods), one country opposed a statement on the risks posed by afforestation and BECCS to biodiversity, water and food security, and livelihoods on the grounds that the IPCC would be contradicting conclusions from SR1.5 on the need for removal. Politically, stating the risks of CDR makes it less desirable, and countries with fossil fuel extraction-based economies have a stake in supporting CDR to continue production. These authors, supported by several other governments, vigorously upheld the scientific basis of their assessment. As a compromise, numerous caveats were added to balance the potential and risks of CDR methods.

So even while a small but vocal contingent within the IPCC has consistently pushed against CDR, it has become a de facto climate solution, one which is both speculative and emerging as a set of technologies, with attendant risks and potential unintended consequences. Other insiders have expressed concern about the IPCC's overreliance on opaque model inputs (Robertson, 2021), but IPCC leadership rebuts them, through a combination of publications, sidelining, editing, and informal conversations to socialize authors into the IPCC's norms (Skea et al., 2021).

Without much fanfare, the IPCC has done a lot of work over the past two assessment cycles to turn a seemingly fabulous, speculative, and risky climate solution into one that appears to be reasonable. Part of this might be because without CDR it is impossible to meet the Paris Agreement temperature goal. AR6 clearly emphasizes the need for CDR unless there is an immediate global transformation of values, energy generation, and the economy. Obviously, we have not yet managed such a transformation, which forces us to move toward climate solutions with significant risks and tradeoffs to limit global warming.

Case Study: Shifting Narratives toward Demand-Side Mitigation

It has been clear for decades that if we are to end global warming, we must change our energy mix from fossil fuels to low- or no-carbon energy sources to stop emitting GHGs. But making that possible relies on social behaviors, policies and political choices, and cultural changes – factors that the IPCC has only hinted at throughout most of its history of assessment. In AR6, these factors were prominently featured, particularly in the WGIII Chapter 5, which assesses demand, services, and the social aspects of mitigation.

While much of WGIII remains entrenched in supply-side economics, Chapter 5 pioneered the assessment of demand-side climate solutions. This chapter was led by strong Chapter Lead Authors (CLAs) who received steady and vocal support from the WGIII co-chairs. One author spoke, at WGIII LAM1, about how Chapter 5's existence is a response to a recurrent critique of the IPCC:

> One thing that I find very interesting is that there is an expectation that the IPCC will be conservative and will suppress, particularly descend around, economic approaches to solving climate. And one of the things I found really surprising, because I'm on Twitter and stuff, was we barely released the 1.5 report ... it wasn't out but half an hour, people definitely read the chapters. They were already saying, you see, there's nothing about de-growth in here. There's nothing ... typical, typical. In all the low-carbon scenarios they're all assuming carbon capture and storage is necessary . . . . They'd really not gone into major economic change. Now I've been frustrated that it has been actually quite hard to get those questions about the literature around major economic change onto the agenda, because it's in the literature, it should be being reviewed. It should be getting a place.

Some IPCC leaders, particularly WGIII co-chair Jim Skea, were vocal about including this in AR6 and successfully lobbied to include the chapter during the AR6 Scoping Meeting, where the reports are outlined and approved by governments. With a totally new chapter, there are no previous models to refer to, so the chapter's narrative and structure must be created fresh. A year later, after the virtual WGIII LAM3, this same interlocutor explained how the chapter's narrative had been developing since the previous interview – they had adopted an "avoid–shift–improve" framework for assessing demand, and connected economic indicators with those of well-being:

> Focusing on demand is really important and that really helps put people into the heart of the whole thing. And that brings in that it's the people's well-being which is really needed, just not the consumption, but the well-being and different indicators for well-being . . . . So in ... the low carbon transition, it's in the societal aspect, and people's need for services. A demand-driven mitigation agenda is one of the most important which IPCC has so far ignored ... [but] we got the chance with this new chapter.

The Chapter 5 authors constructed a narrative that pushed against the dominant frameworks of WGIII, even as those continued to drive much of the report overall. Some authors resisted this on principle, guarding their epistemic vantage. Another noted the extra work Chapter 5 made for the other chapter authors, as they sought inputs in their new-to-IPCC framework. At a coffee break, an author said, "The Demand chapter is the perfect name; they keep making demands of us." Some Chapter 5 authors felt this way too: One expressed, in an interview, excitement and hesitation over their demand chapter's designation as a new mitigation paradigm:

> What is amazing to me is how much the times have changed. And so maybe our chapter sort of played a little bit of a role, but I think also it has been the shift in zeitgeists, or maybe also a growing realization that sort of the economic and political tools that we have for climate change mitigation have not worked particularly well.

Attention to demand-side climate solutions also precipitated the first mention of "sufficiency policies" in an SPM and its definition in a footnote, as "a set of measures and daily practices that avoid demand for energy, materials, land, and water while delivering human wellbeing for all within planetary boundaries." Given that the debate about sufficiency is closely linked to that about de-growth, this was celebrated as a small victory by several authors. An interviewee noted:

> When we got sufficiency into the SPM, I was told by [a Chapter 5 author] "it is more than what we expected" . . . I would have felt awful if the word "sufficiency" had not been included in the SPM. Now that we have the definition, a friend wrote to me "you're celebrating now, but in fact, it's just a footnote."

The writing of ARs, like other knowledge work, is often carried out slowly and deliberately, with innovations tried out inside of – rather than breaking apart – a previously determined structure. Scholars and researchers are generally not disciplined to be revolutionaries even as their research increasingly makes the societal and economic transformations necessary to stop climate change all the more glaring. Chapter 5 provides a stepwise entry into a more transformative assessment, considering economic climate policy from a bottom-up, demand, and behavior-focused perspective. This interrupts the dominant narrative of climate solutions resting solely in national energy policy. Sometimes a subtle mention – like a footnote – can cause a political tsunami, as illustrated for instance by the debate around sufficiency in Europe amid the energy crisis and the publication of AR6. Ultimately, however, while the IPCC reports form a strong knowledge foundation for climate action, they cannot, by design, make the action happen. There are limits to the role that academics take in this problem; others must take the knowledge and move it past the political threshold. "Now do something," an IPCC leader stated as she gaveled an SPM for final approval: It is up to us to act from the assessed climate science.

Feasibility and Enabling Conditions

In AR6, the authors did not only list potential solutions and options but also went on to consider the feasibility – economic, technological, institutional, sociocultural, geophysical, and environmental – of implementing them. While nearly absent in AR5, discussions about feasibility – contextual and subjective as it is – have become a major point of debate both among authors and governments. Here again, SR1.5 played a crucial role in organizing research on feasibility, and the assessment framework that was developed there set the stage for AR6. The IPCC has been an ideal place to kickstart discussions, push disciplinary boundaries, and initiate creative thinking on this topic – and several of the authors in charge of the feasibility assessment in SR1.5 also participated in AR6. As an author told us:

> you really need a community like [that of the IPCC] to make [such assessment] happen. Because [the concept of feasibility] comprises six dimensions and you need a lot of different expertise to do the assessment. I could never do this on my own.

But applying a feasibility framework is not an easy task, because it depends on whether there is relevant literature available, as well as on the willingness of different chapter teams to participate, particularly the sectoral chapters in which energy, transport, buildings, industry, agriculture, forestry, and waste management are assessed. In WGIII, for instance, an author told us there was "some resistance from the sectoral chapters that were not quite comfortable with the methodology." Besides, whether a solution is feasible or not is context-dependent and can change over time. Great care is thus needed when assessing feasibility at a global scale and communicating these results.

It was also not always easy to get authors' assessment of the feasibility of climate solutions accepted by government delegates. A WGII figure (Figure SPM.4) on the feasibility of climate responses and adaptation options was approved, but not without debate. Many countries noted that the figure was too complex. Some delegations also disliked that regional differentiation was missing. Saint Kitts and Nevis, for instance, called for a footnote reflecting regional differences in the feasibility of different options, noting the highly unequal distribution of financial means for adaptation (Earth Negotiations Bulletin, 2022a, p. 13).

WGIII removed from the SPM a similar figure assessing mitigation options after the final government review. In WGIII, it was decided that the figure was too vulnerable: The literature on feasibility was relatively new to the IPCC; the figure had no precedent (a similar figure did not make it into the SPM of SR1.5); the figure had received hostile comments during the governmental review; and, finally, the co-chairs did not feel strongly about the figure. The decision to remove the figure and convey the message in text instead became inevitable, though the figure remained in the Technical Summary (TS.31).

AR6 also went further to spell out the "conditions that enhance such feasibility," called "enabling conditions" – which "include finance, technological innovation, strengthening policy instruments, institutional capacity, multi-level governance and changes in human behaviour and lifestyle" (IPCC, 2022, p. 2907). While existing structures of wealth and power may make some climate solutions easier to achieve than others, the notion of enabling conditions provides an analytical tool for considering how to broaden the reach of possible solutions. Its ethos makes choices seem palatable and reasonable and can set societies up for transformative change in ways that are apparently smooth and easy. It also has the benefit of emphasizing positive factors, by focusing on what can be done, in contrast to approaches focusing on barriers.

AR6 acknowledges that climate solutions do not happen in a vacuum and talks about the multiple social, institutional, and financial conditions that enable their scaling up. This somewhat optimistic tone can hide the nasty political realities that hamper action, which are buried in the reports, but never came close to being mentioned in the SPM.

Conclusions

IPCC solutions are broad, spanning adaptation as well as mitigation, and IPCC leadership continues to expand disciplinary perspectives that can open the "solution space" for their primary audience, policymakers. The solutions the IPCC offers strongly drive how policymakers come to view the menu of possible options, and – as the IPCC also began to frame them – the pathways, the series of choices that lead to particular futures. The reports' narrative style attempts to include a broad set of policy options and the Earth's response to these options. However, not all these solutions are – or even can be – assessed in the same way. The IPCC is biased toward models and quantitative information; some scholarship cannot serve as a model input, nor is all knowledge quantifiable. For example, behavioral and cultural changes cannot be reduced to a piece of code. What moves up into the SPM, though, is often these numerical sets of scenarios and projections, laden with the implicit authority of numbers as facts. This is a serious issue, particularly with IAMs, which attempt to include economic and social choices as numerical codes. This modeling community is, however, much less transparent than the physical earth modeling community whose work anchors WGI.

In WGIII, IPCC solutions, bolstered in large part by opaque IAMs, have been preparing us to meet the 1.5 or 2°C threshold with a narrowing range of options, because our failure to act has made solutions that were previously unfathomable, on both ethical and economic grounds, increasingly necessary. Carbon dioxide removal has now become ingrained into climate solutions. The reports communicate that much more needs to be done, our choices continue to narrow, and the tradeoffs for lowering carbon emissions and holding temperatures as low as possible are increasingly costly, especially in the developing world.

6 Government Approval

While the previous sections briefly mentioned how the Intergovernmental Panel for Climate Change (IPCC)'s Summaries for Policymakers (SPMs) are approved by governments, this section explores this process in greater detail. It shows the enormous amount of work that goes into writing these shorter documents, anticipating their approval, and getting them approved. Such a process often requires very different expertise, skills, and practices than those needed to write the underlying Working Group (WG) reports.

The approval of the SPMs – the short overviews that condense the comprehensive reports by highlighting key policy-relevant messages – is the apex of the assessment process, in which the document is literally and figuratively handed off to governments. Colloquially referred to as simply the "approval," the work of this meeting involves IPCC member governments collectively discussing the SPM and approving its content line by line. While the underlying report – whose content underpins its writing – is accepted without line-by-line scrutiny by governments and remains the responsibility of the authors, the intense political examination to which the SPM is submitted makes it an entirely different sort of document, in form if not in content.

In theory, the idea of producing documents that would be validated by governments seems like a good way to foster co-production between science and politics (Turnhout et al., 2020). The process allows experts to identify and agree on key conclusions that they think are relevant to communicate to inform decisions. Compiled into a single document, these conclusions are then submitted for comments to governments and approved line by line in plenary sessions. In practice, however, it can be an extremely unpleasant, even unhealthy exercise. In the IPCC, this seems to have long been the case. Tony Brenton, a UK diplomat, recalled the first WGI approval session in 1990 as being already chaotic:

> Having started in a very civilized fashion with songs about the future from children's choirs and an address from the prime minister of Sweden, the meeting finally came very close to breakdown. It finished at four o'clock in

the morning, one day late, with most of the delegates having abandoned their chairs in the conference hall to gather on the front podium and shout at each other. (1994, p. 18)

Much of what transpires in an SPM approval is about seeking clarifications and suggesting improvements to the text. It is about reflecting on the tone and emphasis of the SPM and the narratives that it conveys. Much of the meeting time is spent wordsmithing, sometimes at the expense of clarity, to accommodate the numerous requests of governments. It is also through the SPM that new knowledge, scientific terms, and concepts can acquire diplomatic validation and be transmitted to the United Nations Framework Convention on Climate Change (UNFCCC), potentially informing political decisions. Ultimately, getting the document approved is a way of showing the world that the IPCC has, once again, achieved a consensus and that action should follow.

The SPM that results from such negotiations is necessarily a compromise document, one that is expected to reflect the multiple and sometimes contradictory perspectives of the authors, the Bureau members, and the IPCC member states. SPM approval sessions are thus live attempts at co-creating knowledge between those who claim to represent "what the science says" and those who claim to know "what information policymakers need." In practice, the boundary between the two is fuzzy. Even the distinction between authors and delegates is at times confusing: Many delegates have scientific backgrounds, know the literature well, and sometimes contribute to it; and many authors also have an acute sense of what their country's interests are – some of them even have previous work experience in climate negotiations.

The intervention of governments in the writing of the SPMs is important and legitimizes it. Their delegates often have more time to prepare for the approval, while authors juggle their main occupation and their work with the IPCC. Government delegates also look at the SPMs holistically, with an eye toward policymakers' expectations, while authors are responsible for specific parts of the SPM only. Authors generally agree that the approval can help improve the readability and relevance of the SPMs. But not always.

Writing the SPM

The writing of the SPM is a careful and tedious exercise that begins around the time of the second Lead Author Meeting (LAM2). It involves a limited number of authors – about fifty: the Coordinating Lead Authors (CLAs), as well as occasional Lead Authors (LAs) whose expertise cannot be covered by the CLAs, the IPCC Bureau, and the Technical Support Units (TSUs). The writing of the SPM is not only a key phase in the assessment process, during which the report's main findings are laid out, but also an occasion for authors from different chapters to compare their assessments and work on a common narrative. This narrative is conveyed in sections, combining carefully crafted headline statements and paragraphs. The co-chairs generally lead discussions about the overall direction of the SPM and space allocation, while the content is provided by the authors, based on the main conclusions of their chapters.

Each SPM undergoes two main reviews – a government and expert review and a final government review. Between those stages and the approval, the document is significantly modified: Whole sentences are redrafted or deleted, figures are dropped, and new ones are introduced. Several elements are considered in the (re)drafting process: The overall length and readability of the SPM, the robustness and policy-relevance of the statements and visuals, and the traceability of statements and visuals back to the content in the underlying reports. The authors also spend much time discussing the tone of the SPM. In the Sixth Assessment Report (AR6), for instance, many discussions revolved around finding a balance between pessimistic and optimistic messages.

In the SPM editing, the writing team is already anticipating the approval and assessing whether a statement or figure can make it through the process. Self-censorship is common at this stage. Statements and figures that received adverse comments during the review process are likely to be dropped, but the writing team can also decide to defend them in plenary. A great deal of discernment (Oppenheimer et al., 2019) thus comes into play as the writing team needs to find a balance between multiple, sometimes conflicting, comments and concerns. As a TSU member told authors during an SPM writing workshop:

> Reading the governments' comments, I think you can't help but notice that it's slightly schizophrenic. They would like a longer report, a shorter report, something that's less prescriptive, more policy-relevant, we shouldn't go outside our remit, we should, at the same time, speak to the wider issues and a whole bunch of topics. Really, this is *your professional judgment* call as to what you think is the message which best reflects the work you've done on the assessment . . . . (emphasis added)

As the approval comes nearer, the division and compartmentalization of work increases. Discussions take place in breakout groups organized in parallel and few occasions emerge for the writing team to reflect on the report's conclusions as a whole – there is even a tacit expectation not to interfere too much in others' sections or subsections. Much coordination work also goes into allocating individual statements and figures to groups of authors and encouraging them to produce a paper trail documenting and explaining the changes that were made in response to comments. At that stage, the SPM editors are reluctant to introduce new material that has not been run by governments in the review process. In AR6, the WGs organized webinars with governments to gather their last comments, clarify misunderstandings, and probe contentious issues ahead of the approval sessions. They also organized meetings with authors to discuss strategies to respond to governments' interventions and to work on alternative, back-pocket options. Some authors also prepared individually for the approval. "In terms of preparation, I really knew every single sentence, I absolutely knew everything in the chapter . . . , " we were told by a CLA.

There is, however, only so much that can be anticipated. Governments might not have laid all their cards on the table – many do not even partake in the review process, for lack of time, resources, or expertise. And some authors may be ready to defend their conclusions, even though governments might not like them.

The Approval Proceedings

Approvals usually last one week (from Monday to Friday) and take place in a location that can accommodate several hundred people, such as a conference center or hotel. The space always offers a plenary room, as well as several smaller adjacent rooms to schedule parallel meetings, corridors for informal meetings or breaks, and bars and restaurants. These temperature-controlled, homogeneous rooms are open to a plurality of expertise and experience of climate change – though they remain closed to whoever has not gone through a predefined accreditation process. Not all participants, however, arrive at the approval table with the same resources (human, scientific, or diplomatic) and the same level of preparation to weigh in on the discussions.

A SPM approval session usually starts on a slow path with protocolized statements by officials of the host country, the IPCC's parent organizations, and the IPCC Chair. Following the opening session, the work can start. The sections and subsections of the SPM are projected on a large screen and its content – each sentence, footnote, figure, and table – is open to comment. Text is worked on using tracked changes and the status of each sentence is signaled by different colors – yellow when it is under discussion in plenary; green when it is approved; cyan when it is introduced in a context other than the plenary (e.g., a contact group); and purple when it is "parked" – when its consideration is paused and deferred to a later stage. In the absence of requests to take the floor, a gavel is used to mark the end of the discussion.

When the discussion over a statement or a figure is deemed too contentious to be rapidly resolved, it is moved to a parallel meeting or a contact group. Because the negotiations are lengthy – sometimes on purpose – contact groups multiply over the week and the last hours are particularly rushed. When the deliberations extend into the night, the fatigue accumulated over the week puts participants on edge and tensions are palpable. The diplomatic and even enthusiastic tone of the first days gives way to imploring, impatient, and occasionally disrespectful interventions. Remaining compromises are struck in huddles (smaller informal contact groups decided on the spot) or during bilateral exchanges. The prolongation of the deliberations and the multiplication of parallel meetings make it particularly difficult for smaller delegations to attend all discussions.

In AR6, the SPM approval meetings did not unfold as was originally planned. When the COVID-19 pandemic grounded airplanes and dramatically reduced border crossings, the process changed drastically: In 2020, the IPCC decided to postpone the WG approval sessions and to organize them online – for the first time. In comparison to the high ceilings and big halls of the conference centers that usually host its meetings, the institution looked small on individual computers – a screen one saw after clicking a "Join the Meeting" link. Rather than navigating these spaces – the official meeting points, the halls, and restaurants where informal discussions take place – participants navigated Zoom's breakout rooms and struggled with connectivity problems or the needs of children and pets who occasionally entered the perimeter of their screens.

The virtual approval lasted two weeks, twice as long as in-person meetings. Each session had record levels of registered participants – although the discussions were dominated by the same three dozen countries. Authors, government delegates, and observers spent between six and nine hours a day, sometimes more, in front of their computers. In plenary, simultaneous interpretation was provided in all six official UN languages during working hours and an automatic transcription of the conversation was available. The schedule defined by the TSUs (located in France, Germany, and the United Kingdom) and bound to the working hours of the interpreters (based in Switzerland), was in general more amenable for participants in the European and African time zones – despite attempts, as a Bureau member phrased it, "to leave no one behind." Participants who lived elsewhere were more likely to have to work during the night. The usual back-of-the-room discussions between participants were conveyed through instant messaging platforms (WhatsApp or Slack). But communication remained challenging in many respects, especially in the absence of clear channels of communication between the authors and governments.

Approval sessions are never easy to organize but organizing them online was particularly challenging. It was not only a question of making the meetings operational, but also of ensuring that the process was deemed sufficiently inclusive and transparent by governments. This supplementary effort was invisible to the eye of many, and the burden fell particularly hard on the shoulders of the TSUs and the Secretariat. The WGI SPM approval came first. "We are mindful of the fact that we are making IPCC history," claimed Hoesung Lee in his introductory speech on August 26, 2021. The report was particularly anticipated, and it was the only report to be published ahead of 26th Conference of Parties (COP26) to the UNFCCC in Glasgow – as had originally been planned for the entirety of AR6 reports. "The WGI ship continued to sail" despite the global pandemic, noted the WGI co-chair in her welcoming remarks. Impeccably run by the co-chairs, the TSU and the Secretariat, the session finished on time – though not without some last-minute drama.

The WGII SPM approval took place between February 14 and 25, 2022. The deliberations were more contentious, as the Panel dived into the troubled waters of the "solution space." The approval ran over time and the last gavel was heard on Saturday the 26th. The management style was less conciliatory and the tone between the co-chairs and some delegations became heated on several occasions. The work routine was also disturbed by the Russian invasion of Ukraine; the Ukrainian delegates had to leave to take shelter. Speaking from a safe place in the last plenary, the head of the Ukrainian delegation established a connection between climate change and the war:

> ... it's clear that the roots of both these threats to humanity are found in fossil fuels. ... Burning oil, gas and coal is causing warming and impacts we need to adapt to. And Russia sells these resources and uses the money to buy weapons. ... This is a fossil fuel war. (recalled in Milman, 2022)

WGIII approval was scheduled one month later, from March 21 to April 1. The deliberations were extremely repetitive, lengthy, and conflictual – although pressure to accelerate the deliberations was conveyed relatively late in the process. The session finished more than 40 hours later than scheduled after one sleepless night, leaving participants on their last legs. Disagreements were resolved by adding more text, and in particular footnotes – as a result, the size of the SPM increased by one-third. Reflecting on the session, one interviewee noted:

> The circumstances at approval turned out to be at worst. And then some governments just went after us. They got us to add so many qualifiers. ... They were picking on every single line. And I don't think we expected that. We expected some issues to be controversial. ... We didn't expect that it would just go down to the wire and that we would have authors in tears and figures being dropped.

Another author also recalled: "All of this double-guessing what these countries want [ahead of the approval], I just think it's completely unnecessary .... In the end, it didn't work. It was the worst, it was the toughest, most awful thing that ever has happened." Our interviewees pointed to a small number of delegations that were able, over all three-approval sessions, to capitalize on approval practices – which allow governments to take the floor as often as they want and require agreement by all participants – to slow the process and get what they wanted through exhausting others.

The Synthesis Report approval, scheduled one year later (from March 13 to 17, 2023), saw the IPCC going back to in-person approvals in Interlaken (Switzerland), a prospect welcomed by many participants. In practice, however, it quickly became a challenging meeting, as progress was extremely slow, and governments even considered the possibility of having to reconvene at a later stage. The approval was characterized by an unprecedented level of unpreparedness. It concluded 49 hours past its scheduled end, following two sleepless nights. By that time, the representatives of many governments, and particularly those from developing countries, had left, creating a huge backlash.

Global South countries are often accused of slowing down the approval sessions and obstructing the deliberations. While there is no doubt that some countries come to the meetings with the objective of disrupting the process, others put much effort into shaping the SPMs because of the difficulties they face in weighing in on the production of the underlying reports and in enrolling their scientists in the process. As an author reflected,

> The dynamic is that the developed world is better able to shape the report, all the way through the writing or have their perspectives reflected in it .... Developed countries often say [to developing country representatives], why are you speaking so much? Why are you questioning the scientists, etc.? In a sense, it's a reflection of the dynamics of the process, that they have much less ability to shape the report in the front end [the underlying reports], which is why they play a bigger role at the back end [the approval].

The IPCC ship always eventually arrives safely in the harbor and consensus shines out. But there is always a "what if" moment, when participants wonder what would happen if, this time around, IPCC member governments do not reach an agreement. Relief is always palpable when the last gavel of the chair is heard.

Negotiating the SPMs

While hundreds of governments attended the approval sessions, only about forty actively participated. The approvals were particularly technical, with a carefully spelled-out procedure that followed IPCC guidelines as well as diplomatic cultural practices. Government delegates not only sought clarification on the text and visuals that were presented to them, they also questioned whether they accurately reflected the underlying WG reports, the levels of confidence and, sometimes, even the scientific literature. Figures were a major source of discussion, as different stories can emerge from choices around the visual representation of data. Government delegates also questioned what knowledge should be in the SPMs in the first place. They had their own ideas of what policy-relevant information should be introduced and suggested concrete text to elevate to the document. Delegates cannot bring into the SPM information that has not been assessed in the underlying reports. But since the reports are comprehensive and long, there is always plenty of room for interpretation. The approvals were also political. Because IPCC products inform and thereby have the potential to shape international negotiations, in particular within the UNFCCC (see the case studies below), IPCC member governments took a particularly keen interest in the content of the assessments to ensure that their interests were reflected – or at least not undermined.

Any revisions needed to remain true to the authors' understanding of the underpinning scientific knowledge. The reminder was often raised during the approval sessions that authors "hold the pen" and should not be second-guessed. But only a blurry line separates a phrase that reflects the authors' understanding of the underpinning scientific knowledge from one that does not, especially when dealing with conclusions from the social sciences where context matters. Many interventions also concerned issues of emphasis – -what information should come first in the different paragraphs and sections? – and balance – is the text bringing the full picture and reflecting all perspectives, for example, the importance of past, present, and future emissions in determining responsibility to take action, or the opportunities and risks of various options?

Contingent factors also came into play, such as the time a statement or visual was introduced – "at 4 am you just say yes," an author told us – as well as the communication and diplomatic skills of the authors, Bureau members, and delegates. Authors were sometimes pressured to accommodate requests by governments. Juggling between Slack, multiple versions of the SPMs, and the underlying report, they often had to propose changes on the fly. Reflecting, an author told us: "There were one or two points where I thought that we gave away too much, that we actually were pressured by some governments to moderate the conclusions, while I thought there's enough evidence [to stand by them]. So maybe I gave in too much."

Another noted that "it ended up being much more like the more nuanced but less punchy text that I had written to begin with." Huge concessions were made in the last hours of the WGIII and Synthesis Report (SYR) approval sessions. A CLA recalled:

> I kind of really felt [that] what they [some governments] were doing is [that] they were holding everything up to such a degree that they know in the end that they're going to get what they want because everybody's so desperate to get the whole thing signed off. And that's exactly what happened.

SPM Knowledge as Malleable

As has been made clear throughout this Element, writing assessments is not only about getting the facts right but also about conveying narratives. As an author put it, "the SPM approval is a narrative-building process. The approval isn't just about strict scientific knowledge, but how you massage that into a narrative. And politically, it is the narrative that you shape from information that matters the most." This makes knowledge "malleable," and the role of the authors is "to make sure that science shines through, but understanding that people want to see their national context reflected."

Most of the major controversies that arose in the approval process related to the implications the statements and visuals can have for the UNFCCC negotiations and for countries' positions. The two following case studies from WGII and WGIII well illustrate this point. On the one hand, the WGII example reflects the strategy of a few emerging countries to push back against the emphasis given to the 1.5°C target – a strategy also pursued in the UNFCCC (Earth Negotiations Bulletin, 2022a, p. 12). On the other hand, the WGIII example illustrates the challenges the IPCC faces in presenting information that has the potential to reshape countries' understanding of the principle of common but differentiated responsibilities (CBDR) – an issue that already put co-production to a test in AR5 (Dubash et al., 2014).

The two case studies show how the narratives contained in the SPM can be adjusted to respond to governments' interventions. They show the various strategies put in place by governments to shape the SPMs, for example, questioning the authors' understanding of the underlying literature and proposing alternative readings; seeking to weaken the levels of confidence; or, as a last resort, exercising their veto. They also illustrate the authors' and co-chairs' attempts to defend their work and craft compromises, for example, using language to create constructive ambiguities – the way experienced negotiators do – and layers (Biniaz, 2016). The WGIII example shows, however, the limits of co-production in those (rare) moments when knowledge becomes a red flag for countries.

Case Study: 1.5°C

One of the most intense debates in the approval of the WGII SPM concerned section B.6 on the impacts of temporary overshoot, in which the Chinese delegation was opposed to the authors and the co-chairs. An uninformed reader won't find any trace of the controversy in the report, as it is now buried in convoluted phrases and footnotes.

Section B.6 brought new insights about the risks associated with overshooting 1.5°C, thus giving the target a special treatment. Following on the path set by SR1.5, WGII highlighted the benefits of remaining below 1.5°C and put particular emphasis on the impacts of a temporary overshoot on human and nonhuman systems. This emphasis was welcomed by many governments that stressed the policy-relevance of the conclusions, but not all.

The section – comprising one headline statement and three subparagraphs – was introduced on Tuesday, February 22, 2022, in the second week of approval. Discussions were moved to a contact group chaired by Switzerland and Ghana, which met over four days. Early on, the Chinese delegation expressed its reservations, noting that there was, in its view, not enough data on overshoot to support an assessment of its impacts. Other countries asked for clarification on the impact on ecosystems missing from the original statements, and on the conditions for reversibility. Soon, it became obvious that several countries, and particularly China, felt threatened by the explicit association between overshoot and 1.5°C – as opposed to other climate targets, and in particular 2°C, its counterpart in the Paris Agreement (Rajamani and Werksman, 2018). "[Overshooting] is not just about 1.5C," claimed the Chinese delegation. The country's request to remove mention of the 1.5°C threshold in several sentences was, however, opposed by Global North countries and small island states.

As no agreement emerged in the contact group, the statements were brought back to the plenary on Saturday, February 26. The Chinese delegation relentlessly repeated its concerns – poorly conveyed by the interpreters at first – and the disagreement escalated into a tense exchange with the co-chairs. Soon the country stood alone in this fight. Having reached an impasse, it was suggested to follow IPCC procedures, which allow for differing views to be recorded in a footnote (which, in the IPCC process, is perceived as a dreadful threat as it indicates a lack of consensus). Opinions then diverged about whether China should be named in the SPM. It was eventually agreed to convey disagreement through several caveats. The first sentence in B.6.1 now reads: "While model-based assessments of the impacts of overshoot pathways are limited, observations and current understanding of processes permit assessment of impacts from overshoot." In the second sentence, a footnote was added, explaining the choice of a high-confidence statement on the irreversibility of some impacts "despite limited evidence specifically on the impacts of a temporary overshoot of 1.5°C." Lastly, a subtle linguistic trick was used to refer to 1.5°C without overemphasizing it: "Additional warming, *e.g.,* above 1.5°C during an overshoot period this century, will result in irreversible impacts on certain ecosystems with low resilience … " (emphasis added).

The Chinese delegation got its way in paragraph B.6.1. However, its attempts to water down other paragraphs, and particularly the section's headline statement, failed. That statement reads:

> SPM.B.6 If global warming transiently exceeds 1.5°C in the coming decades or later (overshoot)[37], then many human and natural systems will face additional severe risks, compared to remaining below 1.5°C (high confidence). Depending on the magnitude and duration of overshoot, some impacts will cause release of additional greenhouse gases (medium confidence) and some will be irreversible, even if global warming is reduced (high confidence). (Figure SPM.3) {2.5, 3.4, 12.3, 16.6, CCB SLR, CCB DEEP, Box SPM.1} (IPCC 2022a, 19).
>
> [37] In this report, overshoot pathways exceed 1.5°C global warming and then return to that level, or below, after several decades.

Numerous mentions of 1.5°C also remain in other sections – 1.5°C is mentioned 40 times in the approved document, 2°C only 14 times.

Case Study: Categorizing Countries

Another dramatic event was the deletion from the WGIII SPM of a figure (known as Figure SPM.9; see our Figure 5) on realized mitigation investment flows alongside investment needs across sectors, regions, and types of economy. The deletion of a figure is not rare, especially late in the approval process, when time is running out, but it is generally experienced with disappointment and frustration, both by the authors who spent years drafting it, and by the government representatives who lament the failure to find a compromise.

Figure 5 (SPM.9) was introduced on Tuesday, March 22, in the first week of the approval, and moved to a contact group soon after. Dissensus erupted around the presentation of information by type of economy and the differentiation between "developed" and "developing" countries. While the SPM writing team knew the figure was vulnerable – it had received adverse comments during the review process – it had decided to defend it in plenary. For many authors, the developed/developing divide was relevant both scientifically (mentioned at length in the underlying report) and politically (speaking to categories used in the UNFCCC and the Paris Agreement).

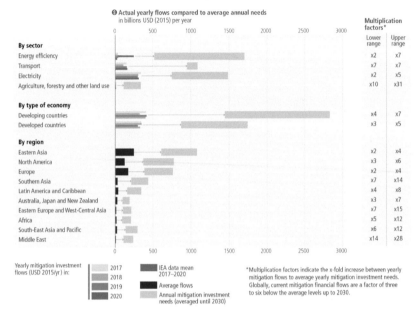

Higher mitigation investment flows required for all sectors and regions to limit global warming

Figure 5 WGIII's deleted Figure SPM.9 (2022b) as published in the SYR (2023), showing mitigation investment needs by sector, type of economy and region. Figure 4.6 in IPCC, 2023: Climate Change 2023: Synthesis Report. Contribution of Working Groups I, II and III to the Sixth Assessment Report of the Intergovernmental Panel on Climate Change [Core Writing Team, H. Lee and J. Romero (eds.)]. IPCC, Geneva, Switzerland, 184 pp., doi: 10.59327/IPCC/AR6-9789291691647.

The use of this country grouping was controversial in other parts of the SPM as well, but the dispute soon crystallized around Figure SPM.9. While the disagreement divided delegations along North/South lines, it developed more specifically into a bilateral confrontation between the USA, for whom such categorization was deemed prejudicial and should be removed, and China, who wanted the differentiation retained. The political stakes are well known and profound: The USA opposes any differentiation and wants all countries, developed and developing, to pursue similar climate action. China wants to maintain its label as a developing country to escape a leading role in climate mitigation, even as the economic realities (and carbon emissions) suggest otherwise.

In the late hour of the approval, the authors tried to save the figure by removing the developed/developing distinction. China, however, vetoed the amendment to the figure, which then remained unacceptable for the USA. Communication at that stage was also difficult. As an interviewee noted,

> In my view, this was also a communication issue. . . . It was so difficult with the translators and not having direct communication. . . . It's so tricky if you cannot get hold of someone. . . . We could not send them [the Chinese] a message and say, "Please be frank and tell us what exactly the issue is."

At a deadlock, the co-chair decided to remove the figure from the SPM. He noted: "We are losing this figure potentially because of four words." Interestingly, the terms developed/developing remained in the text, where they were more acceptable for the USA because of the ambiguity about where the line between developing and developed was drawn. In the figure, however, countries had to be divided explicitly to aggregate investment data by category and China had been grouped among the developing countries. As an author noted:

> We have this funny compromised space where it is ok to use those terms, but not ok in any way to sanctify what goes in each of those terms. The figure was moved out . . . because it risked locking in whose country is developed and whose country is developing, as opposed to the Paris compromise which was self-differentiation.

For several authors, Figure SPM.9 had become the byproduct of an issue that transcended the IPCC. "It had nothing to do with finance," noted an author. "Regionalization and development status is just a huge issue that's beyond IPCC at the moment. And so this issue came crashing into our approval," a Bureau member told us.

The figure features in the WGIII Technical Summary as Figure TS.29 and edited in the Synthesis Report Figure 4.5, though this was not elevated in its SPM. In an unexpected turn of events, a few hours before the closure of the session, the USA accepted the figure, but recorded its opposition to the categorization of developed and developing countries in the report of the session.

Conclusions

We have sought to convey in this section that there is often more to the SPMs than meets the eye. While those not familiar with the IPCC may see a strict display of scientific knowledge that they are happy to take at face value, participants and observers learn to see a web of information woven into a seemingly coherent narrative. For many, narrative-building is a key factor that underpins the policy relevance of the SPMs, in a context in which the documents remain nonbinding. Every bit of statement or visual has a purpose; thus, its rightful place in the SPMs might be contested. It is hoped that, once approved and if put into the right hands, these narratives will find their route out into the world, shape preferences, and inform climate action. The SPMs indeed contain plenty of material that can be used by policymakers nationally and internationally.

Getting their messages across to governments is a challenging experience for many authors, as most of them are not used to acting diplomatically. Appealing to scientific reasoning might work in some cases, but they also need to be ready to adjust messages to reflect the nuances of the world in which we live. This also means sometimes leaving out those messages on which governments cannot agree. To accommodate a variety of national sensibilities in a unique document, the approvals risk reaching "unity in the form of least-common denominator generalities" (Vardy et al., 2017, p. 59). In other cases, because every government needs to get something out of the SPMs and thus put something into it, the different statements of the summary can sometimes cancel each other out. Quotes from the IPCC can be used to call for urgent action and transformative change and, at the same time, for the continuation of fossil fuel exploitation.

Debates about the pros and cons of SPM approvals are brainteasers. On the one hand, the SPMs are important to foster knowledge appropriation by policymakers. As owners of their conclusions, governments can hardly question them – though they can cherry-pick among them and ignore those they do not like. On the other hand, SPM approvals are becoming ever more nightmarish as the IPCC dives deeper into the assessment of climate solutions.

7 How Assessment Practices Shape Climate Knowledge

Inside the IPCC has taken us from the experiences of IPCC authors, to the assessment of climate science, to the integration of risk among Working Groups (WGs), to how IPCC authors consider solutions, and then to the political approval of the Summaries for Policymakers (SPMs). Our research shows the complex, human dimensions of writing assessment reports and the decisions that people must make in their writing – about the science, about representation, about political considerations, and about how to work together. Our sections, in their synthesis, suggest four key arguments.

First, the IPCC borrows practices from science as well as diplomacy throughout its assessment cycle. It also incorporates, with uneven results, diversity, equity, and inclusion practices. However, the IPCC is not an epistemic community in a classical sense, in which authors share a common set of knowledge and work on a problem (scientific and/or political) from that starting point. Instead, the diversity of authors – in every sense – means that there is not a shared set of knowledge beyond a basic adherence to the principles of scientific practice. Authors, instead of serving as a singular epistemic community, serve as advocates for their multiple communities, be they scholarly, geographic, or something else. They bring not only their expertise but also diverse subjectivities that broaden, deepen, and challenge conventional knowledge and practices. They incorporate their known practices for working in groups and diplomatic tactics at various stages in writing the assessment and ushering it through the approval plenaries. Our Element has shown these practices at work and how they ultimately result in a final report that characterizes the state of climate science from a multifaceted global group of experts.

Second, the social or human dimensions of assessment writing are the key strength of the IPCC. Our ethnography has shown a complicated process and some of the tradeoffs that people make to reach agreements. However, this work of putting ideas together by a diverse group of experts is a strength – socially, politically, and even scientifically. Agreements by groups of experts with different knowledge bases, made through writing processes that encourage deliberation, alternate views, and full participation, build trust among authors, in each other and in the assessments they produce. Politically, the work of demonstrating international consensus on climate knowledge matters for international climate policy. At the UNFCCC meetings, the basic science is already agreed upon. Few countries openly question the reality of climate change, but they might disagree about what pieces of the IPCC assessments are most relevant for them. The issues that result in compromise language, even when it weakens statements from the authors, can also demonstrate differentiation and highlight science that is controversial for policymakers, especially when discussing pathways to low-carbon societies. Science has always been a human endeavor – an attempt to discern facts about nature using human intellect and ingenuity. Our ethnography has shown how authors try new tactics for assessing science, leading to new methods, communication strategies, and textual and graphical representations. The institutional environment of the IPCC enables this creativity, even as authors simultaneously experience constraints by governments, their colleagues, or the IPCC Bureau. This is the kind of work that gets approved by governments, that authors are proud of, and that is usable by policymakers and climate advocates.

Third, this Element showed our observations of some of the internal work-ings of IPCC assessment report writing. Coming to agreement about critically important climate science is difficult, and we highlighted some of the chal-lenges authors faced while writing. IPCC authors often perform what Michel DeCerteau (1984), following Claude Lévi-Strauss (1966), calls *bricolage*: The art of sticking different pieces of knowledge and practices together, making do with what is available and improvising when necessary. The final product isn't perfect, and in some ways is quite different from scientific practice, but it is more robust and representative than, say, a report written by a smaller, more homogeneous group. Ultimately, we contend, this makes the report approvable and usable for international climate policy. Disagreements are worked out in the text and figures. Sometimes this results in authors simultaneously feeling unsatisfied about not getting their own perspectives into the chapter and feeling satisfied with the final outcome as a whole. The production of consensus is often the work of least common denominator agreement, but also, as we have shown, it can foster innovation in how to assess and communicate scientific knowledge. This bricolage is know-ledge production in action.

Finally, our study highlights the particular institutional history and culture that shapes procedures and innovation, and, ultimately, how we and our deci-sion-makers understand and respond to climate change. It shows the enormous amount of coordination, cooperation, and integration among authors – and also governments – that brings this report into being. The IPCC as an institution has also become so powerful over time that it drives much of the world's climate knowledge production. That does not mean that the assessment process is all calm and untroubled – indeed, there were instances that caused us alarm during our observations, from harassment accusations to government interventions that sought to muddy the reports and therefore weaken the knowledge they contain. Nonetheless, the most IPCC authors worked tirelessly and took their ethical and scholarly commitments seriously to produce these comprehensive, grave reports about the present and future of Earth's climate decisions and our shrinking opportunities to reverse the course that we have set ourselves upon.

This Element has explored how IPCC authors work together on their assess-ments of climate knowledge as a sociocultural set of processes. The resulting reports are the best set of information we currently have to use as a foundation for debating climate change, from the local to the global. Our Element has shown the successes of and opportunities for improvement for the IPCC. The IPCC itself seems to ascertain this as well, regularly refreshing its leadership, staff, and authors with each assessment report cycle. As we conclude the writing of this Element, AR7 is just beginning to get underway, starting with some new initiatives, a new organizational structure, and new faces in new roles. That being said, efforts will be needed to ensure that lessons from one cycle are not lost in the process of moving to the next.

This Element brought the people behind the IPCC reports to the foreground, showing how the institution works as a community of mostly volunteer climate experts, driven to understand and communicate scientific literature to help decision-makers understand the problem, the scope of the impacts, and a range of solutions and futures available. We also hope that this Element helped translate the IPCC to people interested in climate knowledge and its applications, be it local community climate planning, direct action, or serving as an IPCC author.

References

Allen, M. R., Friedlingstein, P., Girardin, C. A. J., et al. (2022). Net zero: Science, origins, and implications. *Annual Review of Environment and Resources*, **47**(1), 19–39.

Asayama, S., Bellamy, R., Geden, O., Pearce, W. & Hulme, M. (2019). Why setting a climate deadline is dangerous. *Nature Climate Change*, **9**(8), 570–72.

Bassetti, F. (2022). Zeke Hausfather: Every tenth of a degree counts. *Foresight*. October 6. www.climateforesight.eu/interview/zeke-hausfather-every-tenth-of-a-degree-counts.

Beck, S. & Mahony, M. (2018). The politics of anticipation: The IPCC and the negative emissions technologies experience. *Global Sustainability*, **1**(e8), 1–8.

Beck, S. & Oomen, J. (2021). Imagining the corridor of climate mitigation — What is at stake in IPCC's politics of anticipation? *Environmental Science and Policy*, **123**, 169–78.

Betts, R. (2020). Global warming edges closer to Paris Agreement 1.5C limit. *Carbon Brief*. July 9. www.carbonbrief.org/guest-post-global-warming-edges-closer-to-paris-agreement-1-5c-limit.

Biniaz, S. (2016). Comma but differentiated responsibilities: Punctuation and 30 other ways negotiators have resolved issues in the international climate change regime. *Michigan Journal of Environmental & Administrative Law*, **6**(1), 37–63.

Boykoff, M. & Boykoff, J. M. (2004). Balance as bias: Global warming and the US prestige press. *Global Environmental Change*, **14**(2), 125–36.

Boykoff, M. & Pearman, O. (2019). Now or never: How media coverage of the IPCC Special Report on 1.5 C shaped climate-action deadlines. *One Earth*, **1**(3), 285–88.

Brenton, T. (1994). *The Greening of Machiavelli*. London: Earthscan.

Brysse, K., Oreskes, N., O'Reilly, J. & Oppenheimer, M. (2013). Climate change prediction: Erring on the side of least drama? *Global Environmental Change*, **23**(1), 327–37.

Cointe, B. (2022). Scenarios. In K. De Pryck & M. Hulme, eds., *A Critical Assessment of the Intergovernmental Panel on Climate Change*. Cambridge: Cambridge University Press, pp. 137–47.

Cointe, B. & Guillemot, H. (2023). A history of the 1.5 C target. *WIREs Climate Change*, **14**(3), 94-99. https://doi.org/10.1002/wcc.824.

Corbera, E., Calvet-Mir, L., Hughes, H. & Paterson, M. (2016). Patterns of authorship in the IPCC Working Group III report. *Nature Climate Change*, **6**(1), 94–99. https://doi.org10.1038/nclimate2782.

DeCerteau, M. (1984). *The Practice of Everyday Life.* Translated by Steven F. Randall. Berkeley: University of California Press.

De Pryck, K. (2021). Intergovernmental expert consensus in the making: The case of the Summary for Policy Makers of the IPCC 2014 Synthesis Report. *Global Environmental Politics*, **21**(1), 108–29. https://doi.org/10.1162/glep_a_00574.

De Pryck, K. & Hulme, M., eds. (2022). *A Critical Assessment of the Intergovernmental Panel on Climate Change.* Cambridge: Cambridge University Press.

Dubash, N. K., Fleurbaey, M. & Karhta, S. (2014). Political implications of data presentation. *Science*, **345**(6192), 36–37.

Earth Negotiations Bulletin. (2018). Summary of the 48th Session of the Intergovernmental Panel on Climate Change (IPCC–48): 1–6 October. *IISD*, **12**(734), 1–22.

Earth Negotiations Bulletin. (2021). Summary of the 54th Session of the Intergovernmental Panel on Climate Change and the 14th Session of Working Group I: 26 July – 6 August 2021 IISD. 12(781): 1–27.

Earth Negotiations Bulletin. (2022a). Summary of the 55th Session of the Intergovernmental Panel on Climate Change and the 12th Session of Working Group II: 14–27 February. *IISD.* **12**(794): 1–24.

Earth Negotiations Bulletin. (2022b). Summary of the 56th Session of the Intergovernmental Panel on Climate Change and the 14th Session of Working Group III: 21 March–4 April. *IISD.* **12**(795): 1–32.

Geden, O. (2018). Politically informed advice for climate action. *Nature Geoscience*, **11**(6), 380–83.

Geertz, C. (1973). *The Interpretation of Cultures.* New York: Basic Books.

Guillemot, H. (2017). The necessary and inaccessible 1.5 C objective: A turning point in the relations between climate science and politics. In S. Aykut, J. Foyer & E. Morena, eds., *Globalising the Climate: COP21 and the Climatisation of Global Debates.* New York: Routledge, pp. 39–56.

Gunaratnam, Y. (2003). *Researching Race and Ethnicity: Methods, Knowledge and Power.* London: Sage.

Haas, P. M. (1992). Introduction: Epistemic communities and international policy coordination. *International Organization*, **46**(1), 1–35.

Haasnoot, M., Biesbroek, R., Lawrence, J., et al. (2020). Defining the solution space to accelerate climate change adaptation. *Regional Environmental Change*, **20**(2), 1–5.

Hajer, M. & Strengers, B. (2012). Review of: MT Boykoff (2011) Who speaks for the climate? Making sense of media reporting on climate change. *Cambridge Review of International Affairs*, **25**, 298–300.

Haraway, D. (1988). Situated knowledges: The science question in feminism and the privilege of partial perspective. *Feminist Studies*, **14**(3), 575–99.

Harding, S. (1995). "Strong objectivity": A response to the new objectivity question. *Synthese*, **104**, 331–49.

Hausfather, Z. (2018). Analysis: Why the IPCC 1.5C report expanded the carbon budget. *Carbon Brief*. October 8. www.carbonbrief.org/analysis-why-the-ipcc-1-5c-report-expanded-the-carbon-budget/.

Hermanson, L., Smith, D., Seabrook, M., *et al.* (2022). WMO global annual to decadal climate update: A prediction for 2021–25. *Bulletin of the American Meteorological Society*, **103**(4), E1117–29. https://doi.org/10.1175/BAMS-D-20-0311.1.

Hughes, H. (2015). Bourdieu and the IPCC's symbolic power. *Global Environmental Politics*, **15**(4), 85–104.

Hulme, M. (2016). 1.5 C and climate research after the Paris Agreement. *Nature Climate Change*, **6**(3), 222–24.

IPCC. (2018). *Global Warming of 1.5°C: An IPCC Special Report on the Impacts of Global Warming of 1.5°C Above Pre-industrial Levels and Related Global Greenhouse Gas Emission Pathways, in the Context of Strengthening the Global Response to the Threat of Climate Change, Sustainable Development, and Efforts to Eradicate Poverty.* V. Masson-Delmotte, P. Zhai, H.-O. Pörtner, *et al.*, eds. Cambridge: Cambridge University Press.

IPCC. (2020, February). Gender Policy and Implementation Plan. www.ipcc.ch/site/assets/uploads/2020/05/IPCC_Gender_Policy_and_Implementation_Plan.pdf.

IPCC. (2021). *Climate Change 2021: The Physical Science Basis: Contribution of Working Group I to the Sixth Assessment Report of the Intergovernmental Panel on Climate Change.* V. Masson-Delmotte, P. Zhai, A. Pirani, *et al.*, eds. Cambridge: Cambridge University Press.

IPCC. (2022a). *Climate Change 2022: Impacts, Adaptation, and Vulnerability: Contribution of Working Group II to the Sixth Assessment Report of the Intergovernmental Panel on Climate Change.* H.-O. Pörtner, D.C. Roberts, M. Tignor, *et al.*, eds. Cambridge: Cambridge University Press.

IPCC. (2022b). *Climate Change 2022: Mitigation of Climate Change. Contribution of Working Group III to the Sixth Assessment Report of the Intergovernmental Panel on Climate Change.* P. R. Shukla, J. Skea, R. Slade, et al., eds. Cambridge: Cambridge University Press. https://doi.org/10.1017/9781009157926.

IPCC. (2023a). *Climate Change 2023: Synthesis Report.* Contribution of Working Groups I, II and III to the Sixth Assessment Report of the Intergovernmental Panel on Climate Change [Core Writing Team, H. Lee

and J. Romero (eds.)]. IPCC, Geneva, Switzerland, pp. 35–115, https://doi
.org/10.59327/IPCC/AR6-9789291691647.

IPCC. (2023b). IPCC AR6 WGIII: CDR factsheet. www.ipcc.ch/report/ar6/
wg3/downloads/outreach/IPCC_AR6_WGIII_Factsheet_CDR.pdf.

Lahn, B. (2020a). Changing climate change: The carbon budget and the
modifying-work of the IPCC. *Social Studies of Science*, **51**(1), 3–27.

Lahn, B. (2020b). A history of the global carbon budget. *Wiley Interdisciplinary
Reviews: Climate Change*, **11**(3), 1–9.

Latour, B. (1993). *We Have Never Been Modern*. Translated by Catherine
Porter. Cambridge, MA: Harvard University Press.

Lévi-Strauss, C. (1966). *The Savage Mind*. Chicago, IL: University of Chicago
Press.

Liverman, D., vonHedemann, N., Nying'uro, P., et al. (2022). Survey of gender
bias in the IPCC. *Nature*, **602**, 30–32.

Livingston, J. E. & Rummukainen, M. (2020). Taking science by surprise: The
knowledge politics of the IPCC Special Report on 1.5 degrees. *Environmental
Science & Policy*, **112**, 10–16.

Millar, R. J., Fuglestvedt, J. S., Friedlingstein, P., et al. (2017). Emission
budgets and pathways consistent with limiting warming to 1.5 C. *Nature
Geoscience*, **10**(10), 741–47.

Milman, O. (2022). "This is a fossil fuel war": Ukraine's top climate scientist
speaks out. *The Guardian*, www.theguardian.com/environment/2022/mar/
09/ukraine-climate-scientist-russia-invasion-fossil-fuels.

O'Brien, K., Eriksen, S., Nygaard, L. P. & Schjolden, A. (2007). Why different
interpretations of vulnerability matter in climate change discourses. *Climate
Policy*, **7**(1), 73–88.

O'Neill, S. J., Hulme, M., Turnpenny, J. & Screen, J. A. (2010). Disciplines,
geography, and gender in the framing of climate change. *Bulletin of the
American Meteorological Society*, **91**(8), 997–1002. www.jstor.org/stable/
26232948.

Oppenheimer, M., Oreskes, N. & Jamieson, D., et al. (2019). *Discerning
Experts: The Practices of Scientific Assessment for Environmental Policy*.
Chicago, IL: The University of Chicago Press.

O'Reilly, J. (2017). *The Technocratic Antarctic: An Ethnography of Scientific
Expertise and Environmental Governance*. Cornell, NY: Cornell University
Press.

O'Reilly, J., Oreskes, N. & Oppenheimer, M. (2012). The rapid disintegration of
projections: The West Antarctic Ice Sheet and the Intergovernmental Panel
on Climate Change. *Social Studies of Science*, **42**(5), 709–31.

Oreskes, N. & Conway, E. M. (2011). *Merchants of Doubt*. New York: Bloomsbury. https://doi/10.1017/CBO9781107415324.004.

Pelling, M., High, C., Dearing, J. & Smith, D. (2008). Shadow spaces for social learning: A relational understanding of adaptive capacity to climate change within organisations. *Environment and Planning A*, **40**(4), 867–84.

Pickering, A. (2009). The politics of theory. *Journal of Cultural Economy*, **2**(1–2), 197–212.

Plumer, B. & Davenport, C. (2019). Science under attack: How Trump is sidelining researchers and their work. *The New York Times*. December 28.

Rajamani, L. & Werksman, J. (2018). The legal character and operational relevance of the Paris Agreement's temperature goal. *Philosophical Transactions of the Royal Society A*, **376**(2119), 1–14. https://doi.org/10.1098/rsta.2016.0458.

Randalls, S. (2010). History of the 2°C climate target. *WIREs Climate Change*, **1**, 598–605. https://doi.org/10.1002/wcc.62.

Robertson, S. (2021). Transparency, trust, and integrated assessment models: An ethical consideration for the Intergovernmental Panel on Climate Change. *Wiley Interdisciplinary Reviews: Climate Change*, **12**(1), 1–8.

Schipper, E. L. F. (2020). Maladaptation: When adaptation to climate change goes very wrong. *One Earth*, **3**(4), 409–14.

Shapin, S. (2010). *Never Pure: Historical Studies of Science as if It Was Produced by People with Bodies, Situated in Time, Space, Culture, and Society, and Struggling for Credibility and Authority*. Baltimore, MD: JHU Press.

Shove, E., Pantzar, M. & Watson, M. (2012). *The Dynamics of Social Practice: Everyday Life and How It Changes*. Thousand Oaks, CA: Sage.

Simpson, N. P., Mach, K. J., Constable, A., *et al.* (2021). A framework for complex climate change risk assessment. *One Earth*, **4**(4), 489–501.

Sismondo, S. (2010). *An Introduction to Science and Technology Studies*. Chichester: Wiley-Blackwell.

Skea, J., Shukla, P., Al Khourdajie, A. & McCollum, D. (2021). Intergovernmental Panel on Climate Change: Transparency and integrated assessment modeling. *Wiley Interdisciplinary Reviews: Climate Change*, **12**(5), 1–11.

Standring, A. & Lidskog, R. (2021). (How) Does diversity still matter for the IPCC? Instrumental, substantive and co-productive logics of diversity in global environmental assessments. *Climate*, **9**(6), 1–15.

Sundqvist, G., Bohlin, I., Hermansen, E. A. & Yearley, S. (2015). Formalization and separation: A systematic basis for interpreting approaches to summarizing science for climate policy. *Social Studies of Science*, **45**(3), 416–40.

Tandon, A. (2023). Analysis: How the diversity of IPCC authors has changed over three decades. *Carbon Brief*. www.carbonbrief.org/analysis-how-the-diversity-of-ipcc-authors-has-changed-over-three-decades/.

Taylor, M. (2014). *The Political Ecology of Climate Change Adaptation: Livelihoods, Agrarian Change and the Conflicts of Development*. New York: Routledge.

Turnhout, E., Metze T. A. P., Wyborn, C., Klenk, N. & Louder, E. (2020). The politics of co-production: Participation, power, and transformation. *Current Opinion in Environmental Sustainability*, **42**, 15–21. https://doi/10.1016/j.cosust.2019.11.009.

United Nations. (2021). *Secretary-General Calls Latest IPCC Climate Report "Code Red for Humanity," Stressing "Irrefutable" Evidence of Human Influence* [press release]. November 15. https://press.un.org/en/2021/sgsm20847.doc.htm.

United Nations. (2022a). IPCC adaptation report "a damning indictment of failed global leadership on climate." *UN News*. February 28. https://news.un.org/en/story/2022/02/1112852.

United Nations. (2022b). *Secretary-General Warns of Climate Emergency, Calling Intergovernmental Panel's Report "a File of Shame," While Saying Leaders "Are Lying," Fuelling Flames* [press release]. April 4. https://press.un.org/en/2022/sgsm21228.doc.htm.

Van Coppenolle, H., Blondeel, M. & Van de Graaf, T. (2022). Reframing the climate debate: The origins and diffusion of net zero pledges. *Global Policy*, **14**(1), 1–13. https://doi/10.1111/1758-5899.13161.

Vardy, M. (2022). Integration. In K. De Pryck & M. Hulme, eds., *A Critical Assessment of the Intergovernmental Panel on Climate Change*. Cambridge: Cambridge University Press, pp. 169–77.

Vardy, M., Oppenheimer, M., Dubash, N. K., O'Reilly, J. & Jamieson, D. (2017). The intergovernmental panel on climate change: Challenges and opportunities. *Annual Review of Environment and Resources*, **42**, 55–75.

Venturini, T., De Pryck, K. & Ackland, R. (2023). Bridging in network organisations: The case of the Intergovernmental Panel on Climate Change (IPCC). *Social Networks*, **75**, 137–47. https://doi/10.1016/j.socnet.2022.01.015.

Acknowledgments

We thank the following people for their intellectual and logistical support of our project: Barbara Breitung, Mike Hulme, Dale Jamieson, Michael Oppenheimer, Naomi Oreskes, and Franck Petiteville. We thank David Konisky and Aseem Prakash for their support of our manuscript, Megan Pugh, Keynyn Brysse, the editors at Cambridge University Press, and our reviewers. We appreciate institutional support from International Studies, Hamilton Lugar School at Indiana University Bloomington, Department of Geography at the University of Cambridge, Sciences Po Grenoble, the University of Geneva, and Kwantlen Polytechnic University. We thank our IU student researchers Victoria Lincourt, Kari Peiscop-Grau, Thomas Day, and Sarah Meadows, support by Indiana University's Integrated Program in the Environment and the Tobias Center for Innovation in International Development. We also thank our families and friends for their support and encouragement over the years of this research.

This project was funded by the National Science Foundation, grant #1643524; the Swiss National Science Foundation (SNSF), grants #P2GEP1_188326 and P400PS_199235; the Climate Futures Initiative at Princeton University; the Center for Energy and Environmental Research in the Human Sciences at Rice University; and the 0.6 percent Faculty PD Fund at Kwantlen Polytechnic University. We thank them for their support.

Finally, we thank IPCC authors, IPCC bureau members, and the WGI, WGII, and WGIII Technical Support Units for helping us navigate AR6 and for generously sharing their experiences with us. We hope that our Element helps shed light on the work they contribute to helping us understand and solve climate change.

Cambridge Elements ≡

Organizational Response to Climate Change

Aseem Prakash

University of Washington

Aseem Prakash is Professor of Political Science, the Walker Family Professor for the College of Arts and Sciences, and the Founding Director of the Center for Environmental Politics at University of Washington, Seattle. His recent awards include the American Political Science Association's 2020 Elinor Ostrom Career Achievement Award in recognition of "lifetime contribution to the study of science, technology, and environmental politics," the International Studies Association's 2019 Distinguished International Political Economy Scholar Award that recognizes "outstanding senior scholars whose influence and path-breaking intellectual work will continue to impact the field for years to come," and the European Consortium for Political Research Standing Group on Regulatory Governance's 2018 Regulatory Studies Development Award that recognizes a senior scholar who has made notable "contributions to the field of regulatory governance."

Jennifer Hadden

University of Maryland

Jennifer Hadden is Associate Professor in the Department of Government and Politics at the University of Maryland. She conducts research in international relations, environmental politics, network analysis, nonstate actors, and social movements. Her research has been published in various journals, including the *British Journal of Political Science, International Studies Quarterly, Global Environmental Politics, Environmental Politics,* and *Mobilization.* Dr. Hadden's award-winning book, *Networks in Contention: The Divisive Politics of Global Climate Change,* was published by Cambridge University Press in 2015. Her research has been supported by a Fulbright Fellowship, as well as grants from the National Science Foundation, the National Socio-Environmental Synthesis Center, and others. She held an International Affairs Fellowship from the Council on Foreign Relations for the 2015–16 academic year, supporting work on the Paris Climate Conference in the Office of the Special Envoy for Climate Change at the US Department of State.

David Konisky

Indiana University

David Konisky is Professor at the Paul H. O'Neill School of Public and Environmental Affairs, Indiana University, Bloomington. His research focuses on US environmental and energy policy, with particular emphasis on regulation, federalism and state politics, public opinion, and environmental justice. His research has been published in various journals, including the *American Journal of Political Science, Climatic Change,* the *Journal of Politics, Nature Energy,* and *Public Opinion Quarterly.* He has authored or edited six books on environmental politics and policy, including *Fifty Years at the U.S. Environmental Protection Agency: Progress, Retrenchment and Opportunities* (Rowman & Littlefield, 2020, with Jim Barnes and John D. Graham), *Failed Promises: Evaluating the Federal Government's Response to Environmental Justice* (MIT, 2015), and *Cheap and Clean: How Americans Think about Energy in the Age of Global Warming* (MIT, 2014, with Steve Ansolabehere). Konisky's research has been funded by the National Science Foundation, the Russell Sage Foundation, and the Alfred P. Sloan Foundation. Konisky is currently coeditor of *Environmental Politics.*

Matthew Potoski

UC Santa Barbara

Matthew Potoski is a Professor at UCSB's Bren School of Environmental Science and Management. He currently teaches courses on corporate environmental management, and his research focuses on management, voluntary environmental programs, and public policy. His research has appeared in business journals such as *Strategic Management Journal, Business Strategy and the Environment,* and the *Journal of Cleaner Production,* as well as public policy and management journals such as *Public Administration Review* and the *Journal of Policy Analysis and Management.* He coauthored *The Voluntary Environmentalists* (Cambridge, 2006) and *Complex Contracting* (Cambridge, 2014; the winner of the 2014 Best Book Award, American Society for Public Administration, Section on Public Administration Research) and was coeditor of *Voluntary Programs* (MIT, 2009). Professor Potoski is currently coeditor of the *Journal of Policy Analysis and Management* and the *International Public Management Journal.*

About the Series

How are governments, businesses, and nonprofits responding to the climate challenge in terms of what they do, how they function, and how they govern themselves? This series seeks to understand why and how they make these choices and with what consequence for the organization and the eco-system within which it functions.

Cambridge Elements $\equiv$

Organizational Response to Climate Change

Printed in the United States
by Baker & Taylor Publisher Services